Biochar

Fueling a New Economy in the Developing World

Editors

Sara de Jesus Duarte
Latin America Institute of Biochar, México

Patrick Atanasije Pineda
Latin America Institute of Biochar, México

CRC Press
Taylor & Francis Group
Boca Raton London New York

CRC Press is an imprint of the
Taylor & Francis Group, an **informa** business
A SCIENCE PUBLISHERS BOOK

Cover credit: Cover image and section figures reproduced by kind permission of Ivan Moises Gutierrez.

First edition published 2026
by CRC Press
2385 NW Executive Center Drive, Suite 320, Boca Raton FL 33431

and by CRC Press
4 Park Square, Milton Park, Abingdon, Oxon, OX14 4RN

CRC Press is an imprint of Taylor & Francis Group, LLC

Library of Congress Cataloging-in-Publication Data (applied for)

ISBN: 978-1-032-72178-1 (hbk)
ISBN: 978-1-032-74164-2 (pbk)
ISBN: 978-1-003-46794-6 (ebk)

DOI: 10.1201/9781003467946

Typeset in Times New Roman
by Prime Publishing Services

Acknowledgments

We extend our deepest gratitude to the numerous contributors who wrote these book chapters, providing expert opinions that ensured the high technical and scientific quality of this book. In particular, we thank our internationally recognized team: Dr. Bruno Glaser, Dr. Arthur Gruss, Dr. Leônidas Carrijo Azevedo Melo, Dr. Wenceslau Geraldes Teixeira, Dr. Frank Verheijen, Dr. Edson Márcio Mattiello, Dr. José Ferreira Lustosa Filho, Dr. Cícero Célio de Figueiredo, Dr. Deborah Aller, Dr. Martha Lustosa Carvalho, Ines Serra Baucells, Daniel Velasquez, Albert Bates, Dr. Suzanne Allaire, Dr. Shanthi Prabha Viswanathan, Bluvin Ravindran, Kathleen Draper, Dr. Stephen Joseph, and Thomas Miles, as well as several anonymous reviewers and interviewees whose contributions were invaluable.

We are indebted to Ivan Moises Gutierrez, who created our cover and section figures. Sincere thanks go to Raju, the editor at Editorial Department Publications Ltd, for his expert guidance throughout the publication process, and to the Taylor and Francis team for refining our chapters. We also acknowledge the financial support from "The Next 150" and the dedicated assistance of Luisa Marin, Ives Ranver, and Guadalupe Avendaño.

Finally, we thank our families and friends for their patience during the intensive process of bringing this book to life, their understanding through countless late nights of writing, and their unwavering support, without which this work would not have been possible.

Preface

The transformative potential of biochar for economic development and climate change mitigation has now come to light, revealing opportunities that reach from individual entrepreneurs to national economies. If you want to discover how biochar can create new economic pathways while supporting a sustainable future, then it is with great excitement that we share this book: *Biochar: Fueling a New Economy in the Developing World.*

Written for entrepreneurs, project developers, policymakers, farmers, and biochar enthusiasts, for anyone seeking to make a meaningful impact through science, innovation, and sustainability. Our approach combines peer-reviewed research, and industry insights to provide both vision and practical pathways for action.

We offer a broader and bolder vision of biochar as an enabler for financial resilience, livelihood support, climate change mitigation and its multiple opportunities.

Now it's time to join the journey. Learn, innovate, act—and together, let's unlock the full potential of biochar to change lives, communities, and economies in the developing world.

This book is organized into three main sections, each guiding the reader through a different dimension of biochar's transformative potential:

1. **Biochar: Evolution, Adoption, and Agricultural Applications** – This section traces the historical development of biochar, beginning with its origins in the ancient Amazonian Terra Preta and extending to its modern advancements in technology and industrial applications. It follows by examining how biochar enhances agriculture through targeted blending strategies, reduces reliance on chemical fertilizers and fosters sustainable soil management. Building on this foundation, it emphasizes the importance of promoting biochar education in developing countries, with a focus on strategies for widespread adoption.
2. **Biochar: Unlocking Applications and Carbon Credit Pathways** – This section widens the perspective beyond agriculture, showcasing biochar's diverse applications. In carbon credits framework, it presents evidence

demonstrating why biochar is a more effective tool for generating carbon credits compared to other amendments. This section also provides a practical guide—from feasibility assessments to credit issuance—designed as an essential resource for transforming concepts into impactful, financeable ventures.

3. **Shaping Tomorrow with Biochar: Policies, Education, and Real-World Insights** – The final section connects science with society, demonstrating how knowledge-sharing, policy frameworks, and global case studies drive biochar adoption. Readers will encounter examples of entrepreneurs, companies, and governments already transforming challenges into opportunities, offering a roadmap for scaling biochar impact in developing contexts.

Sara de Jesus Duarte
Patrick Pineda

Contents

Section 3 – Shaping Tomorrow with Biochar: Policies, Education, and Real-World Insights

Section 1 – Biochar: Evolution, Adoption, and Agricultural Applications

Terra Preta was first observed in 1542. Early hypotheses suggested that its remarkable fertility arose from the careful, prolonged deposition of organic kitchen waste, charred plant material, and other residues by indigenous Amazonian communities. Today, biochar—the vital component of Terra Preta—stands as a scalable, cost-efficient climate solution, offering the remarkable ability to remove carbon while delivering broad agricultural and industrial benefits. Yet, despite this extraordinary potential, many people remain unaware of what biochar is. By fostering education, raising awareness, and supporting policies, especially in developing countries, we can open the door to its widespread adoption. In agriculture, biochar holds tremendous promise, particularly when combined with other soil amendments, helping to nourish the land and support sustainable farming for generations to come.

KEY LESSONS

- Never underestimate the power of ancient wisdom; it can inspire solutions to the world's greatest challenges, climate change and food security.
- Spreading knowledge about biochar through education is essential to promote supportive policies, drive adoption, and foster climate-smart agriculture in developing countries.
- Embracing biochar is a key step toward thoughtful soil management, offering a safe way to reduce chemical fertilizer use, enhance crop yields, benefiting both people and the planet.

CHAPTER
1

Biochar: From Terra Preta Discovery to Modern Industry

Bruno Glaser* **and Arthur Gross**
Martin Luther University Halle-Wittenberg, Institute of Agricultural and Nutritional Sciences, Soil Biogeochemistry, Germany

1. Introduction

Biochar, a carbon-rich material produced through the thermal decomposition of organic matter under limited oxygen conditions, has gained significant attention for its potential to enhance soil fertility, sequester carbon, and mitigate environmental challenges. Its origins can be traced back to ancient Amazonian civilizations, where indigenous communities created Terra Preta—highly fertile dark earth soils—through the accumulation of biochar, organic residues, and nutrient-rich materials. These anthropogenic soils remain a testament to biochar's long-term stability and effectiveness in improving soil productivity.

The resurgence of interest in biochar stems from its ability to address pressing global challenges, including soil degradation, food security, and climate change mitigation. Scientific research over the past two decades has demonstrated biochar's capacity to improve soil structure, enhance microbial activity and diversity, increase water retention, and reduce greenhouse gas emissions. This has led to its adoption in various research fields beyond agriculture, including environmental remediation, waste management, and industrial applications.

Despite these promising findings, biochar research is still evolving, with ongoing debates regarding its long-term stability, variability based on feedstock and production conditions, and potential risks in large-scale applications. Additionally, the commercialization and widespread adoption of biochar face economic and policy-related challenges that require further exploration.

*Corresponding author: bruno.glaser@landw.uni-halle.de

DOI: 10.1201/9781003467946-2

This chapter aims to provide a comprehensive review of biochar's historical development, characteristics, and modern applications. It will explore its role in soil amendment and environmental management, outline key milestones in biochar research, and examine technological advancements that have expanded its industrial and commercial uses. By bridging traditional knowledge with contemporary scientific findings, this chapter highlights biochar's potential as a sustainable tool for addressing global environmental and agricultural challenges.

2. Importance of Biochar for Terra Preta and other Anthrosol Formation

Biochar has an ancient lineage rooted in traditional practices of indigenous Amazonian cultures. This chapter traces the journey of biochar from the creation of fertile Terra Preta soils to its development as a versatile modern environmental solution. The focus includes biochar's early agricultural applications, its defining characteristics, and how technological advances have expanded their role in sustainable agriculture and environmental remediation.

Biochar is a form of pyrogenic organic matter (PyOM), produced via thermochemical conversion (typically pyrolysis or gasification) of organic material under limited oxygen supply with the purpose of soil amelioration. This process results in polyaromatic carbon compounds, which remain in the soil for centuries, enhancing soil carbon sequestration and soil resilience against climate change (Gross et al. 2021, 2024; Kuzyakov et al. 2014). The high carbon content of biochar includes condensed aromatic structures, which contribute to its persistence and effectiveness as a soil amendment. Biochar also increases soil cation exchange capacity (CEC), especially in long-aged samples (Ke et al. 2022).

The concept of biochar as a carbon-rich product intended for soil amelioration was first recognized in peer-reviewed literature in the early 2000s (Glaser et al. 2002; Lehmann et al. 2006). A profound peer-reviewed reference is Lehmann and Joseph (2009), in which biochar is described in detail as a tool for carbon sequestration and soil improvement, distinguishing it as a unique material with specific environmental benefits, unlike traditional charcoal. This comprehensive peer-reviewed textbook compiled the state-of-the art of research on biochar and its applications in soil amelioration, synthesizing findings from earlier studies and setting the stage for the subsequent research focus on biochar as a soil amendment. In 2024, the 3rd edition of this biochar compendium was published (Lehmann and Joseph, 2024).

Terra Preta was first observed by Spanish explorer Francisco de Orellana in 1542 during his Amazon expedition, where he noted densely populated riverbanks (Glaser 2021b). Early hypotheses proposed that Terra Preta's fertility resulted from prolonged deposition of organic kitchen waste, charred plant material, and other residues by indigenous Amazonian communities (Glaser 2021b). Meanwhile, several natural soil formation processes were proposed such as alluvial and volcanic depositions, but also man-made plaggen depositions (Glaser 2021b). It took until the 1960s for researchers, led by Dutch soil scientist Wim Sombroek, and later by the German soil scientist Wolfgang Zech, to come back to the man-made hypothesis of Terra Preta formation. Studies by Glaser et al. (2001) and Glaser and Birk (2012) finally confirmed the man-made origin of Terra Preta using sophisticated soil biogeochemical analysis including black carbon and steroid analyses.

Biochar, referred to as "pyrogenic carbon" or "black carbon", is crucial in Terra Preta's formation (Glaser et al. 2001). The biochar found in these soils results from the *in-situ* burning of organic materials, including wood, plant residues, and bones, under low-oxygen conditions, leading to highly stable carbon compounds. Biochar in Terra Preta is characterized by its condensed aromatic structures, allowing it to remain stable in the soil for thousands of years. This stability enhances soil fertility and nutrient retention, making Terra Preta remarkably fertile compared to surrounding soils (Glaser et al. 2001).

Terra Preta soils belong to a series of highly fertile and C-rich Anthrosols called "Anthropogenic Dark Earths". While they differ in their origin and age, they all share the same key ingredient being biochar (Kern et al. 2019). Soil organic matter of Terra Preta in South America is composed of about 35% biochar (Glaser et al. 2001). Terra Preta soils contain high levels of stable organic matter, including approximately 200 tons of organic carbon plus 50 tons biochar per hectare in the top meter (Glaser et al. 2001). These soils contain 70 times more biochar than surrounding soils, making biochar the key ingredient of Terra Preta besides tremendous amounts of nutrients and microbial stabilization (Glaser and Birk 2012). The biochar in Terra Preta acts as a long-term carbon sink and retains essential nutrients such as nitrogen and phosphorus (Glaser 2007).

Found in medieval and ancient European settlements, European Dark Earths are enriched with biochar, organic refuse, and ashes, and a similar origin to Terra Preta was concluded (Wiedner et al. 2015). Created through similar methods as Terra Preta, African Dark Earths show high levels of biochar and nutrients, transforming nutrient-poor tropical soils into highly fertile ground (Solomon et al. 2016). In many areas, paring-and-burning practices, which involve burning grassland or heathland sods, provide immediate nutrient availability (especially phosphorus and potassium) and improve soil drainage. These systems yield pseudo-sand from fused clay particles, which enhances soil structure and prevents waterlogging (McKey 2020).

Microbial communities, particularly fungi, played a crucial role in stabilizing biochar and forming stable soil organic matter within Terra Preta. Contrary to myths suggesting a unique "Terra Preta code", research shows that these microbial communities are different in different Terra Preta soils but also differ in Terra Preta and surrounding soils (Glaser 2021a). Modern research utilizes biomarkers to identify specific organic and microbial residues in Terra Preta and verify the contribution of biochar and other organic materials as key ingredients to Terra Preta formation. Methods like molecular markers and isotope analysis and electron microscopy have identified various sources of nutrients, such as bones (for calcium and phosphorus) and animal excrement (for nitrogen and other elements), underscoring the role of human waste management practices in Terra Preta's enrichment (Glaser and Birk 2012).

The Terra Preta phenomenon showcases biochar's ability to transform poor soils into productive, nutrient-rich lands, with implications for modern sustainable practices. The presence of biochar, combined with organic waste, creates a stable and fertile soil ecosystem, highlighting a valuable approach to soil management that can be replicated globally. The Terra Preta model suggests that biochar can be a sustainable way to improve soil fertility, sequester carbon, and recycle organic waste. This concept is inspiring modern biochar applications in agriculture and waste management, particularly in regions with nutrient-poor soils.

3. Characteristics and Functions of Biochar

Biochar's evolution from Terra Preta soils to modern applications was driven by its unique properties and technological advancements in production. The structure of biochar, highly porous and carbon-dense, makes it ideal for retaining nutrients and water, supporting its use across agricultural, environmental, and industrial applications.

Biochar properties are highly dependent on the pyrolysis condition and the feedstock (Schimmelpfennig and Glaser 2012). The key pyrolysis condition affecting biochar properties is the temperature used for pyrolysis but also includes the heating rate, the residence time, and the purging gas used. Biochar properties can also be altered by different modification methods that are applied before or after pyrolysis. The widely variable biochar types (and the resulting properties) and the large number of potential applications mean that results from biochar research are often highly variable (Schmidt et al. 2021).

Biochar increases nutrient retention, soil water-holding capacity, and promotes microbial activity (Schmidt et al. 2021). This leads to improved soil structure, enhanced plant growth, and reduced leaching of essential nutrients. Biochar-enriched soils support diverse microbial communities. The high porosity and surface area of biochar enhance aeration, water retention, and soil aggregation, which are particularly beneficial in sandy and degraded soils. Biochar is known for its high carbon stability, porosity, and ability to adsorb nutrients and pollutants. These properties depend significantly on the feedstock material and pyrolysis conditions, including temperature and residence time (Schimmelpfennig and Glaser 2012). At lower temperatures, biochar is more porous and nutrient-absorbent, while higher temperatures yield more stable forms suitable for long-term carbon storage. Biochar's carbon-rich composition sequesters carbon in soil, reducing atmospheric CO_2 and contributing to climate change mitigation. Its ability to bind nutrients and contaminants also offers potential in pollution control (Gross et al. 2021).

4. Chronology of Biochar Research

The first peer-reviewed articles in the ISI Web of Science database using the search-term "biochar" date back to between 1998 and 2004 and deal with the use of biochar as renewable fuels to substitute fossil fuels. Glaser et al. (2002) provided the earliest review on using biochar for amelioration of chemical and physical soil properties (Table 1). Five papers published in 2007 dealt with biochar as a soil amendment (Chan et al. 2007; Rondon et al. 2007; Warnock et al. 2007), or as a carbon-negative renewable energy source (Lehmann 2007a,

Table 1. Overview of major developments in biochar research, from foundational studies to recent advancements in application methods and environmental impact

Publication year	Key development in biochar research	Reference
2002	One of the earliest studies on using biochar for amelioration of chemical and physical soil properties, using the term "charcoal" at that time	(Glaser et al. 2002)
2005–2010	Significant increase in biochar publications, with focus on biochar for soil productivity and carbon sequestration. Early conferences and collaborations emerged, supporting widespread study	(Lehmann and Joseph 2009; Verheijen et al. 2010)
2010	Acceleration in biochar research topics includes soil improvement, climate change mitigation, and pollutant sorption applications. First meta-analysis on biochar effects on plant growth	(Jeffery et al. 2011)
2012	Release of European Biochar Certificate standardizing research methodologies, production, and control	(Schmidt et al. 2020)
2014	Comprehensive meta-analyses on biochar impacts on soil properties and crop yields, with emphasis on agronomic and environmental benefits, especially in less-developed regions	(Cayuela et al. 2014)
2012–2016	EU Cost Action on biochar coordinating European Biochar research, bringing together researchers, stakeholders and potential users from EU and candidate countries	(Tammeorg et al. 2017)
2017	Biochar studies expanded to nanoscale biochar particles, highlighting their unique properties and greater efficiency in environmental applications	(Archanjo et al. 2017)
2021	First meta-analysis of meta-analyses	(Schmidt et al. 2021)
2024	3rd Edition of most comprehensive biochar book for environmental management	(Lehmann and Joseph 2024)

2007b). Those pioneering work strongly promoted biochar research. Eventually, the number of publications on biochar exponentially increased to more than 23,000 by the end of 2024 in ISI Web of Science (Fig. 1). Most of those publications are focused on biochar's agronomic and environmental benefits, with "biochar and global climate change" and "soil quality and plant growth" as some of the main research topics (Schmidt et al. 2021). It appears noteworthy that there are almost four times more publications on "biochar" compared to the research topic "soil organic matter" (Fig. 1).

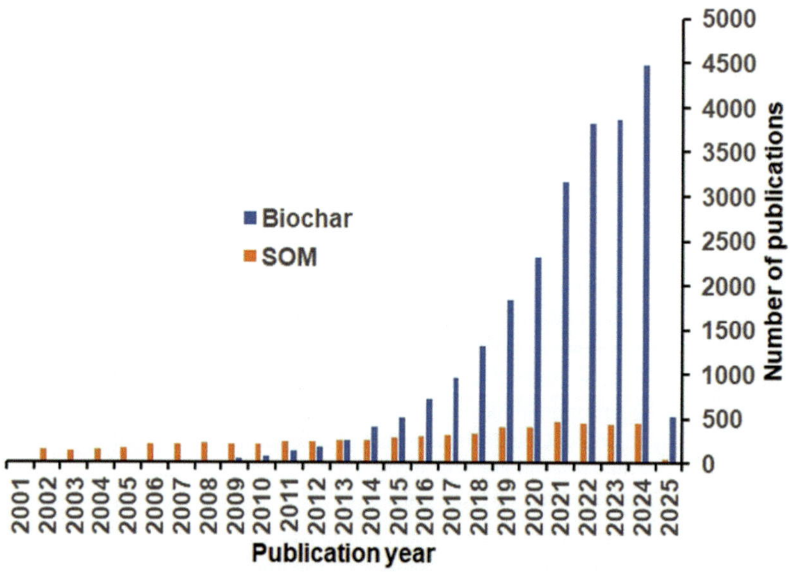

Figure 1. Chronology of biochar publications in comparison to soil organic matter (SOM) using the terms "biochar" and "soil organic matter" as title in ISI Web of Science (accessed on 16.02.2025).

5. Chronology of Biochar Technology and Industry

While ancient applications of biochar were limited to agricultural enhancement, recent decades have seen a broader recognition of biochar's environmental benefits (Table 2). Studies indicate that biochar reduces greenhouse gas emissions, supports waste management, and improves soil and water quality, marking it a multi-functional tool for modern environmental challenges (Glaser et al. 2024). Historically, biochar-like materials have been utilized in diverse regions, including Asia, Africa, and Oceania, where societies incorporated biochar into soil for long-term fertility enhancement. In modern times, biochar is employed in carbon markets, construction materials, and water filtration, underscoring its role in a sustainable economy (Glaser et al. 2024). Technological advancements

Table 2. Overview of major developments in biochar technology and industry

Year	Key development in biochar technology and industry	Reference
Ancient	Early biochar production through traditional methods for soil improvement and crop yield enhancement, utilized in various cultures for centuries	(Glaser et al. 2024)
2000	First biochar field trial in Central Amazonia	(Birk et al. 2009)
2009	First designer biochar described	(Novak et al. 2009)
2009	First commercial large-scale biochar production technology deployed	Www.pyreg.de
2012	First commercial biochar production facility in Europe	Www.sonnenerde.at
2012	55 ideas for the sustainable use of biochar	(Schmidt 2012)
2015	Integration of biochar into environmental management, used increasingly for water purification, air pollutant adsorption, and soil conditioning applications	(Lehmann and Stephen 2015)
2016	Technological advances included optimized pyrolysis and activation processes, creating biochar with increased surface area and functionality for various industrial applications	(Bachmann et al. 2016)
2017	Expanded biochar applications in agriculture and environmental remediation, with significant interest in biochar as a renewable and sustainable material in building and chemical industries	(Maroušek et al. 2017)
2018	Development of engineered biochars with tailored properties for specific applications in pollution control, catalysis, and energy storage, addressing industrial and environmental needs	(Tareq et al. 2019)
2019	Biochar commercialization initiatives launched, moving toward circular economy applications, with standards developed for consistent quality and environmental safety	(Wu et al. 2019)
2020	Biochar applications diversified, including advancements in smart biochar for wastewater treatment, solid fuel alternatives, and adsorbents in hazardous waste management	(Xiang et al. 2020)
2021	Large-scale biochar production technology refined for greater cost-effectiveness and scalability, allowing it to become a more viable industry-standard material	(Hu et al. 2021)
2024	Biochar is the biggest carbon dioxide removal (CDR) technology, which is not governmentally regulated	www.cdr.fyi

have refined biochar production, enabling controlled pyrolysis from diverse feedstocks, including agricultural and urban waste materials. This flexibility not only aids waste management but also creates a carbon sink through carbon sequestration, as biochar's stability means it resists decomposition for centuries (Paradelo et al. 2024).

This timeline highlights the evolution of biochar technology from traditional practices to advanced industrial applications across environmental and agricultural sectors. Each step illustrates biochar's progression towards greater scalability, environmental impact, and economic potential. Biochar's transformation from a traditional soil amendment to a modern environmental tool highlights its versatility and potential. Today, biochar bridges historical practices with innovative applications, addressing global issues in sustainable agriculture, carbon sequestration, and pollution management.

After its primary application in soil enhancement, biochar has found significant utilization in several key markets such as agriculture, animal farming, industrial applications and the carbon dioxide removal (CDR) market: Dominating the biochar market, agriculture accounts for a substantial share of biochar applications. In 2023, the agricultural sector held over 77% of the market revenue, primarily due to biochar's benefits in improving soil fertility, water retention, and crop yields. Biochar is increasingly used in animal farming as a feed additive. Its inclusion in livestock diets has been linked to improved digestion and reduced methane emissions, contributing to more sustainable farming practices. Beyond agriculture, biochar is utilized in various industrial sectors. Notably, it serves as an effective material for water treatment, leveraging its porous structure to remove contaminants. Additionally, biochar is explored for energy storage applications, such as in supercapacitors, and as an eco-friendly component in construction materials.

6. Future Perspectives for Biochar Use and Market Expansion

The biochar market is poised for substantial growth in the coming years. Valued at approximately USD 541.8 million in 2023, the market is projected to reach USD 1.35 billion by 2030, with a compound annual growth rate (CAGR) of 13.9%. Biochar plays a significant role in the CDR market due to its ability to sequester carbon effectively and provide additional environmental benefits. Biochar is produced through the pyrolysis of organic biomass, a process that stabilizes the carbon absorbed by plants during photosynthesis, converting it into a form that can persist in soils for hundreds to thousands of years. This long-term sequestration makes biochar a potent tool for carbon removal. Beyond carbon storage, biochar enhances soil health, improves water retention, and increases agricultural productivity, offering a multifaceted approach to environmental sustainability.

As of mid-2023, biochar-based carbon removal accounted for approximately 92% of all durable CDR deliveries, highlighting its dominance in the carbon removal sector. The global biochar market has experienced rapid growth, with revenues soaring from $600 million in 2023 to a projected $3.3 billion by 2025. This expansion is driven by increasing demand for sustainable carbon removal solutions and the recognition of biochar's co-benefits in agriculture and environmental management.

Biochar presents a cost-effective option for carbon removal. The average cost of biochar carbon removal is $179 per ton of CO_2, significantly lower than the average price of $388 per ton across all CDR methodologies. This affordability enhances its attractiveness to both investors and policymakers seeking efficient carbon offset strategies.

The biochar industry is on a trajectory to deliver a gigaton of carbon removal by 2040, marking a substantial contribution to global climate mitigation efforts. This growth is supported by technological advancements, policy incentives, and increasing corporate commitments to carbon neutrality. For instance, in January 2025, Google announced an agreement to purchase 100,000 tons of carbon removal credits from Varaha, an Indian company specializing in converting agricultural waste into biochar. Such investments underscore the expanding role of biochar in corporate sustainability strategies and the broader CDR market.

7. Conclusions and Recommendations

In summary, the biochar market is experiencing significant growth, driven by its versatile applications in agriculture, animal farming, and industry. With increasing emphasis on sustainability and technological progress, biochar is set to play a crucial role in future environmental management and industrial practices. In addition, biochar stands out in the CDR market for its effectiveness in long-term carbon sequestration, cost efficiency, and additional environmental benefits. Its growing adoption and market expansion reflects its potential as a pivotal tool in addressing climate change challenges.

References

Archanjo, B.S., Mendoza, M.E., Albu, M., Mitchell, D.R.G., Hagemann, N., Mayrhofer, C. et al. 2017. Nanoscale analyses of the surface structure and composition of biochars extracted from field trials or after co-composting using advanced analytical electron microscopy. *Geoderma* 294: 70–79.

Bachmann, H.J., Bucheli, T.D., Dieguez-Alonso, A., Fabbri, D., Knicker, H., Schmidt, H.P. et al. 2016. Toward the standardization of biochar analysis: The COST Action TD1107 interlaboratory comparison. *Journal of Agricultural and Food Chemistry* 64(2): 513–527.

Birk, J.J., Steiner, C., Teixiera, W.C., Zech, W., and Glaser, B. 2009. Microbial response to charcoal amendments and fertilization of a highly weathered tropical soil. *In: Amazonian Dark Earths: Wim Sombroek's Vision*, 309–324.

Cayuela, M.L., van Zwieten, L., Singh, B.P., Jeffery, S., Roig, A., and Sánchez-Monedero, M.A. 2014. Biochar's role in mitigating soil nitrous oxide emissions: A review and meta-analysis. *Agriculture, Ecosystems and Environment* 191: 5–16.

Chan, K.Y., Van Zwieten, L., Meszaros, I., Downie, A., and Joseph, S. 2007. Agronomic values of green waste biochar as a soil amendment. *Australian Journal of Soil Research* 45(8): 629–634.

Glaser, B., Haumaier, L., Guggenberger, G., and Zech, W. 2001. The "Terra Preta" phenomenon: A model for sustainable agriculture in the humid tropics. *Naturwissenschaften* 88(1): 37–41.

Glaser, B., Lehmann, J., Steiner, C., Nehls, T., Yousaf, M., and Zech, W. 2002. Potential of pyrolyzed organic matter in soil amelioration. *People's Republic of China Ministry of Water Resources 12th International Soil Conservation Organization Conference*, Beijing, China, 421–427.

Glaser, B., Lehmann, J., and Zech, W. 2002. Ameliorating physical and chemical properties of highly weathered soils in the tropics with charcoal – A review. *Biology and Fertility of Soils* 35(4): 219–230.

Glaser, B. 2007. Prehistorically modified soils of central Amazonia: A model for sustainable agriculture in the twenty-first century. *Philosophical Transactions of the Royal Society B: Biological Sciences* 362(1478): 187–196.

Glaser, B., and J.J. Birk. 2012. State of the scientific knowledge on properties and genesis of Anthropogenic Dark Earths in Central Amazonia (terra preta de índio). *Geochimica et Cosmochimica Acta* 82: 39–51.

Glaser, B. 2021a. Terra Preta – Fakten und Mythen. *Biologie in Unserer Zeit* 51(4): 360–367.

Glaser, B. 2021b. Terra Preta – Entstehung und Rolle für Klimaschutz und Stoffkreisläufe. *In:* J.L. Lozán [ed.], *Warnsignal Klima: Boden and Landnutzung*, 380–387.

Glaser, B., McKey, D., Giani, L., Teixeira, W., di Rauso Simeone, G., Schneeweiß, J. et al. 2024. Historical accumulation of biochar as a soil amendment. *In: Biochar for Environmental Management*, 15–56.

Gross, A., Bromm, T., Polifka, S., Fischer, D., and Glaser, B. 2024. Long-term biochar and soil organic carbon stability—Evidence from long-term field experiments in Germany. *Science of the Total Environment* 954(June): 176340.

Hu, Q., Jung, J., Chen, D., Leong, K., Song, S., Li, F. et al. 2021. Biochar industry to circular economy. *Science of the Total Environment* 757: 143820.

Jeffery, S., Verheijen, F.G.A., van der Velde, M. and Bastos, A.C. 2011. A quantitative review of the effects of biochar application to soils on crop productivity using meta-analysis. *Agriculture, Ecosystems and Environment* 144(1): 175–187.

Kuzyakov, Y., Bogomolova, I., and Glaser, B. 2014. Biochar stability in soil: Decomposition during eight years and transformation as assessed by compound-specific 14C analysis. *Soil Biology and Biochemistry* 70: 229–236.

Lehmann, J., Gaunt, J., and Rondon, M. 2006. Bio-char sequestration in terrestrial ecosystems – A review. *Mitigation and Adaptation Strategies for Global Change* 11(2): 395–419.

Lehmann, J. 2007a. A handful of carbon. *Nature* 447(7141): 143–144.

Lehmann, J. 2007b. Bio-energy in the black. *Frontiers in Ecology and the Environment* 5(7): 381–387.

Lehmann, J., and Joseph, S. 2009. Biochar for environmental management – An introduction. *In: Biochar for Environmental Management*, 1–18.

Lehmann, J., and S. Joseph. 2024. *Biochar for Environmental Management*. Science, Technology and Implementation Third Edition. ISBN: 978-1-032-28615-0

Schmidt, H.-P. 2012. 55 Uses of biochar. *Ithaka Journal* 25(1/2012): 13–25.

Schmidt, H.-P., Kammann, C., and Hagemann, N. 2020. *EBC-Guidelines for the Certification of Biochar-Based Carbon Sinks*, Version 2.1.

Warnock, D.D., Lehmann, J., Kuyper, T.W., and Rillig, M.C. 2007. Mycorrhizal responses to biochar in soil – Concepts and mechanisms. *Plant and Soil* 300(1–2): 9–20.

Wiedner, K., Schneeweiß, J., Dippold, M.A., and Glaser, B. 2015. Anthropogenic Dark Earth in Northern Germany – The Nordic analogue to Terra Preta de Índio in Amazonia. *Catena* 132: 114–125.

Wu, P., Ata-Ul-Karim, S.T., Singh, B.P., Wang, H., Wu, T., Liu, C. et al. 2019. A scientometric review of biochar research in the past 20 years (1998–2018). *Biochar* 1(1): 23–43.

Xiang, W., Zhang, X., Chen, J., Zou, W., He, F., Hu, X. et al. 2020. Biochar technology in wastewater treatment: A critical review. *Chemosphere* 252: 126539.

CHAPTER

2

Biochar and Biochar-based Fertilizer: Enhancing Soil and Reducing Fertilizer and Water Usage

Leônidas Carrijo Azevedo Melo[1*], Wenceslau Geraldes Teixeira[2], Frank Verheijen[3], Edson Márcio Mattiello[4], José Ferreira Lustosa Filho[5] and Cícero Célio de Figueiredo[5]

[1] Department of Soil Science, Federal University of Lavras - UFLA, Lavras, MG, 37202-203, Brazil

[2] National Soil Research Center, Embrapa Solos, 22460-000, Rio de Janeiro, RJ, Brazil

[3] Centre for Environmental and Marine Studies (CESAM), Department of Environment and Planning, University of Aveiro, Campus Universitario de Santiago, 3810-193 Aveiro, Portugal

[4] Hub-P-CEFENP, Department of Soil Science, Universidade Federal de Viçosa, Av. Peter Henry Rolfs, s/n, 36570-900, Viçosa, MG, Brazil

[5] Faculty of Agronomy and Veterinary Medicine, University of Brasília, 70910-970, Brasília, DF, Brazil

1. Introduction

Soil fertility is fundamental to sustainable agriculture but maintaining nutrient availability while minimizing environmental impact remains a challenge. In this context, biochar, a carbon-rich material produced through pyrolysis of organic biomass, has gained attention as a soil amendment with multiple agronomic and environmental benefits. Its application has been shown to improve soil structure, enhance nutrient retention, and increase fertilizer use efficiency, contributing to both crop productivity and sustainability.

Due to its high porosity, biochar improves soil aeration and water retention, which is particularly beneficial in drought-prone regions. Additionally, its capacity to adsorb and retain nutrients helps prevent nutrient leaching, enhancing the efficiency of nitrogen (N), phosphorus (P), and potassium (K) fertilizers. Some studies suggest that biochar application has the potential to reduce the need for

*Corresponding author: leonidas.melo@ufla.br

DOI: 10.1201/9781003467946-3

chemical fertilizers by improving nutrient availability and uptake, which could lead to higher yields with lower inputs.

Beyond its role in soil fertility, biochar also contributes to climate change mitigation by sequestering carbon in the soil and reducing nitrous oxide (N_2O) emissions associated with nitrogen fertilization. However, despite its multiple advantages, large-scale adoption remains a challenge due to the relatively high application rates required for optimal results.

This chapter explores the effects of biochar on soil fertility and fertilizer savings, highlighting its potential to enhance nutrient use efficiency, improve soil health, and reduce fertilizer dependency. Additionally, it examines how biochar can be used to develop enhanced fertilizers, specifically biochar-based fertilizers (BBFs), which aim to reduce the need for high application rates while improving the economic feasibility of large-scale biochar use. Finally, the chapter discusses how biochar influences soil water-holding capacity and the mechanisms underlying this effect. By integrating biochar into nutrient management strategies, agriculture can move toward more sustainable and efficient soil management practices.

2. Biochar Effects on Soil Fertility and Fertilizer Saving

Biochar, a carbon-rich material produced through the pyrolysis of organic biomass under limited oxygen conditions, has emerged as a promising solution for sustainable soil management (Wu et al. 2019). Biochar can be produced from a wide range of feedstocks, which helps reduce environmental impact, improve soil health, enhance crop productivity, and mitigates climate change (Ding et al. 2016). Biochar offers a range of benefits, including enhanced soil aeration and water retention due to its porous nature, which allows for better root growth and improves the soil's ability to retain water, making it particularly valuable in drought-prone regions. By increasing the soil's water-holding capacity, biochar helps crops maintain a more consistent water supply, reducing stress and promoting healthier plant growth (Razzaghi et al. 2020).

In addition to improving soil structure, biochar plays a crucial role in nutrient retention and enhanced soil fertility. Nutrient-rich feedstocks such as coffee husk generates high cation exchange capacity (CEC) biochar, which rapidly translate this CEC once applied into soil (Domingues et al. 2020). Biochar aging in soil can also contribute to increase soil CEC and nutrient retention, which allows it to hold onto essential nutrients such as nitrogen and potassium, preventing them from leaching into groundwater (Hossain et al. 2020; Wang et al. 2020). Over time, biochar contributes to the build-up of soil organic carbon (SOC), which is vital for maintaining soil fertility and productivity (Hossain et al. 2020).

As biochar application to soil has potential to improve soil properties over time, it is expected that biochar might reduce the need for frequent fertilizer applications and also improve nutrient use efficiency. Tian et al. (2022) found that applying 10 t/ha of biochar significantly improved rapeseed growth, nitrogen use efficiency (NUE), and grain yield. When combined with reduced nitrogen rates of 180 kg/ha and 144 kg/ha, NUE increased by 10%, and grain yield rose by up to 21.8%. The study also showed that biochar compensated for reduced nitrogen fertilizer, maintaining high productivity while promoting root development and enhancing nitrogen uptake. Additionally, biochar had lasting effects on soil health, with benefits persisting for at least two years. The combination of B10 (10 t/ha) with N80 (80% of recommended N rate) was identified as the most efficient, suggesting biochar can reduce nitrogen fertilizer use by 20% while maintaining high yields and sustainability in rapeseed cultivation. Ning et al. (2022) examined the effects of N fertilizer reduction combined with 15 t/ha of rice straw biochar on rice yield, soil fertility, and N use efficiency in a two-year field experiment. The study found that reducing N alone slightly decreased yield, but adding biochar improved rice tiller numbers, biomass, and grain yield by 5.8% with 20% N reduction and 7.7% with 40% N reduction. Biochar also enhanced N use efficiency, increasing agronomic N efficiency by up to 309% and N recovery by up to 151%. Additionally, biochar improved soil pH and organic matter. Overall, combining biochar with reduced N fertilizer is an effective strategy for maintaining rice yields and improving soil health.

Zhang et al. (2024) showed that combining biochar with water-saving irrigation (WSI) improves rice root growth, nitrogen absorption, and yield. Four treatments were tested, with varying levels of rice straw biochar application and irrigation techniques, comparing them to traditional flooded irrigation as a control. Optimal biochar application (12.5 t ha^{-1}) increased nitrogen use efficiency (NUE) by 39.4% and nitrogen agronomic efficiency (NAE) by 24.5%. This strategy-maintained rice productivity while enhancing soil health, making it a sustainable approach for rice cultivation in high-fertility soils of Northeast China. For P, Qi et al. (2023) examined the effects of combining biochar and WSI on P dynamics, rice yield, and sustainability. Biochar (20–40 t/ha) reduced P leaching by 6.77–17.6%, while controlled irrigation decreased total P losses by 35.2–55.1%. Biochar also improved P use efficiency by 5.3–10.3% and increased rice yields by 17–34%, leading to annual net profits of 358–652 USD ha^{-1} y^{-1}. The study recommends applying 40 t/ha biochar with WSI for sustainable rice production and better environmental outcomes.

The application of nitrogen fertilizers to soils contributes to climate change by increasing N_2O emissions through intensified denitrification processes. In this context, biochar serves as a valuable tool to mitigate these emissions, as it not only sequesters carbon in the soil but also reduces N_2O emissions (Cayuela et al. 2014). Grutzmacher et al. (2018) reported a significant reduction in greenhouse gas emissions in soils amended with biochar compared to both unamended and feedstock-treated soils. Specifically, they observed a 72% reduction in CO_2 emissions in biochar-amended soil compared to feedstock-amended soil and up to an 87% reduction in N_2O emissions.

Beyond its benefits to soil health and climate, biochar also plays a significant role in reducing environmental pollution. It helps prevent nutrient runoff and leaching, major causes of water pollution and eutrophication. By retaining nutrients within the soil, biochar reduces the risk of contaminating nearby water bodies, which is especially important in areas where excessive fertilizer use leads to environmental degradation. Laird et al. (2010) observed that biochar significantly reduces nutrient leaching in agricultural soils, as demonstrated in a study using Midwestern soils amended with biochar and swine manure. The application of biochar at increasing rates reduced the total N and P leached from soil columns by 11% and 69%, respectively. This reduction in nutrient leaching is attributed to biochar's ability to retain nutrients through physical, chemical, and biological processes. The study suggests that biochar amendments can effectively limit nutrient runoff, enhance nutrient retention and reducing environmental contamination of water bodies.

Biochar holds great potential for enhancing soil fertility by improving nutrient retention and use efficiency, which can lead to significant fertilizer savings. Integrating biochar into soil management practices presents a sustainable solution for reducing fertilizer dependence while maintaining, or even increasing, crop productivity. However, despite these advantages, large-scale adoption remains challenging due to the relatively high application rates required for optimal results. A promising approach to address this limitation is combining biochar with fertilizers, which can lower the required biochar application rate while still achieving substantial yield improvements as discussed in the next topic.

3. Biochar-based Fertilizers

The production of biochar-based fertilizers (BBFs) is emerging as an alternative to enhance the use of biochar as both a soil amendment and a nutrient carrier for agricultural applications. The development of BBFs is driven by two primary strategies: (i) enriching biochar with one or more nutrients that are either in low concentrations or unbalanced in its composition, and (ii) leveraging the inherent properties of biochar, particularly its negative surface charges and high porosity, to improve nutrient use efficiency while reducing the potential for contamination associated with conventional fertilizers. In general, BBFs are produced with the purpose of reducing the release rate of highly soluble fertilizer nutrients (Ndoung et al. 2023), reducing nutrient losses through volatilization (Rizhiya et al. 2020) and leaching (Fachini et al. 2022), increasing carbon sequestration (Dominguez et al. 2020), and reducing the availability of contaminants.

The most commonly used biomasses for producing BBFs include sludges (such as sewage sludge and paper fiber sludge), crop residues (e.g. straw, corncob, stover, and husk), poultry litter, manure, and tree prunings. The feedstocks employed to enrich or impregnate biochar in the production of BBFs can be categorized into five groups: (i) conventional fertilizers, (ii) rock dust, (iii) minerals, (iv) pure inorganic compounds, and (v) organic wastes.

3.1 Synthesis of Biochar-based Fertilizers

Mixing feedstocks to produce BBFs can be performed in pre-, post-pyrolysis, or combining these processes (Figure 1). The mixtures can also undergo post-pyrolysis treatment including impregnation, enrichment or encapsulation. BBFs can remain powder or be subjected to granulation or pelletization processes.

In pre-pyrolysis, biochar enrichment can be achieved through direct treatment or pre-treatment. Direct treatment involves mixing nutrient-rich feedstocks such as sludges, algae, compost, mushroom substrate, and manures, along with specific nutrient-rich materials like P-rich bone waste or K-rich banana peduncles. The

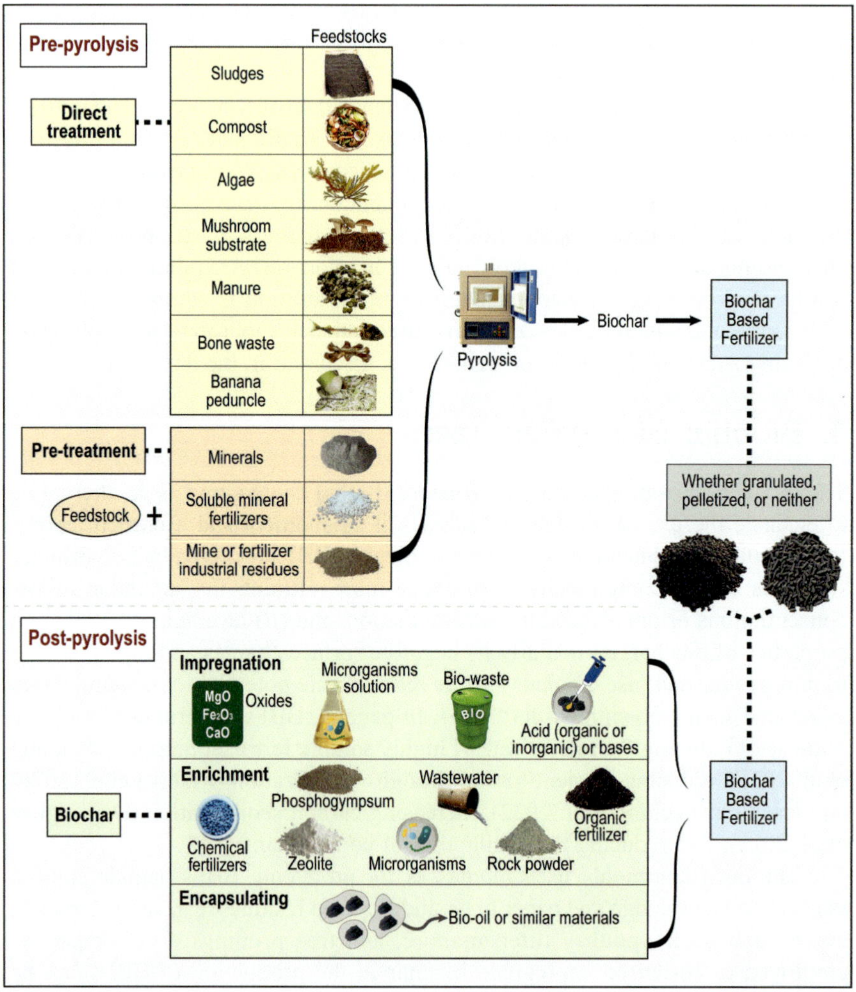

Figure 1. Schematic diagram illustrating routes for biochar-based fertilizer synthesis.

mixture is subjected to slow pyrolysis, potentially followed by post-pyrolysis treatment. Alternatively, nutrient-poor biomass can be mixed with nutrient-rich materials during pre-pyrolysis. This technique typically utilizes minerals (e.g. goethite, dolomite, gypsum, diatomite, bentonite, zeolite, and glauconite), soluble mineral fertilizers (like urea and triple superphosphate), and waste from the fertilizer and mining industries. For example, a slow-release, high-efficiency P BBF was produced by mixing coffee husk and chicken litter enriched with phosphoric acid and magnesium oxide combined with triple superphosphate (Carneiro et al. 2021). Similarly, coffee husk mixed with glauconite activated with potassium hydroxide before pyrolysis resulted in increased total potassium content and availability in the soil (Leite et al. 2024). When biochar is enriched with metals during pre-pyrolysis, particularly iron and zinc, its physicochemical properties are substantially modified. Co-pyrolysis of eucalyptus biomass with $ZnCl_2$ significantly enhances biochar formation, especially with respect to properties ideal for adsorbent materials. Zinc-enriched biochar exhibited a threefold increase in specific surface area (940 m^2/g) compared to conventional biochar. Additionally, its total pore volume and average pore radius increased fivefold, further boosting its adsorption potential (Almeida Leite et al. 2023).

The most common BBF production process involves mixing biochar with nutrient-rich materials, which can be impregnated with solutions or encapsulated to enhance fertilizer performance. Biochar enrichment is the primary method for BBF production, using conventional soluble fertilizers (such as urea, ammonium sulfate, triple superphosphate, single superphosphate, diammonium phosphate, and potassium chloride), along with calcium and magnesium oxides, phosphogypsum, rock powder, and zeolites. Post-pyrolysis enrichment aims to increase nutrient concentration, improve nutrient use efficiency, and reduce gaseous losses, contamination, and leaching. For instance, BBF created by enriching sewage sludge biochar with KCl achieved slow-release fertilizer criteria and reduced K leaching losses by 70% compared to conventional KCl (Fachini et al. 2022). Additionally, organic fertilizers can be incorporated in post-pyrolysis BBF production; for example, a BBF was produced by mixing biochar, rock phosphate, and cow manure in ratios of 45%, 35%, and 20%, respectively (Moradi et al. 2019).

Enrichment with organic fertilizer enhances nutrient availability and promotes the activity of specific soil microorganisms. BBFs can influence soil microorganisms by altering the abundance, activity, and diversity of the soil microbial community. For instance, the application of BBFs has been shown to increase the abundance of growth-promoting bacteria in rice plants (Chew et al. 2020). Similarly, Zhou et al. (2019) observed an increase in microbial biomass and bacterial diversity in calcareous soils following the addition of BBFs. The high porosity of BBFs facilitates the colonization of microorganisms. Consequently, the interaction between BBFs and microorganisms improves soil chemical properties. Adding BBFs can enhance nutrient content, cation exchange capacity (CEC), pH, and water-holding capacity, thereby indirectly influencing the diversity and structure of the soil microbial community. Moradi et al. (2019) reported an increased abundance of nitrifying bacteria in saline soils treated with BBF, thereby enhancing nitrogen availability for plants. In addition to direct enrichment with nutrient-rich sources, BBFs can be produced by impregnating biochar with organic or inorganic acids, bases, oxides, microorganisms, or biowastes. They can also be encapsulated with bio-oil or similar materials to enhance their slow-release nutrient properties. Encapsulated BBFs provide a gradual nutrient release, improve soil physical, chemical, and biological properties, and immobilize heavy metals as the coating degrades (Sim et al. 2021). Furthermore, BBFs can undergo granulation or pelletization, commonly applied in post-pyrolysis techniques.

3.2 Biochar-based Fertilizers Impact on Soil Physicochemical Properties

In recent years, BBFs have emerged as a transformative approach in sustainable agriculture, enhancing nutrient use efficiency, soil health, and reducing reliance on conventional fertilizers. Studies show that BBFs often maintain or increase productivity levels compared to traditional fertilization methods, underscoring their potential in modern agronomy.

One key benefit of BBFs is their positive impact on soil chemical and physical properties. Their application has been linked to improved soil aeration and increased water retention capacity. For example, biochar-based urea fertilizers elevated ammonium and water-soluble organic carbon concentrations, enhancing aeration and microbial dynamics, as evidenced by a higher pmoA/mcrA ratio (Zhou et al. 2024). This improvement in soil structure supports root development and nutrient uptake efficiency. Additionally, incorporating materials like hydrotalcite and starch into BBFs significantly enhances soil water retention, providing a resilient approach to farming in drought-prone regions (Lu et al. 2023).

BBFs also enhance nutrient use efficiency. A study by Morais and Silva (2023) found that BBFs made from coffee husk-derived biochar and acidulated apatite increased soil P availability by 43% after maize cultivation compared to mineral fertilizers (TSP), promoting a 46% increase in P accumulation in maize shoots. Fachini et al. (2024) demonstrated that K-enriched BBFs improved nutrient release synchronization, leading to greater tuber production in radish plants.

Furthermore, adding biochar to fertilizers significantly reduces N losses from leaching and volatilization, enhancing nitrogen use efficiency. Biochar-based urea fertilization reduced CO_2 and N_2O emissions from subtropical soils in China compared to standard urea fertilization (Zhang et al. 2023). If 20% of the region's forest area (4.68 million ha) switched from urea to biochar-based urea, potential annual reductions in CO_2 and N_2O emissions could reach 2,740 t CO_2 yr^{-1} and 667 t N_2O yr^{-1}, respectively.

The alkaline nature of certain BBFs can also increase soil pH. BBFs produced by co-pyrolysis of poultry litter with phosphate and magnesium raised soil pH after maize cultivation compared to TSP (Lustosa Filho et al. 2017). This increase in pH improves the soil's ability to retain nutrients and reduces heavy metal toxicity and bioavailability. BBFs made from biogas residue and magnesium ammonium phosphate also elevated soil pH and decreased heavy metal pollution risks associated with conventional fertilizers (Luo et al. 2024). However, the liming effect of some BBFs can increase the pH of acidic soils and contribute substantially to higher rates of NH_3 volatilization. For example, in a field experiment in southeastern China, the application of biochar-based urea (BBU) significantly increased NH_3 volatilization due to the characteristics of increasing soil pH after applying BBU. However, BBU treatment reduced NO_3^--N leaching losses from 53.3% to 33.4% compared to conventional fertilizer (urea) (Zhou et al. 2021). Additionally, it should be highlighted that leaching was the main N loss pathway in all the treatments, accounting for more than 95% of N losses from N fertilizer. In another study with BBU, soil pH slightly increased due to BBU application, and concentrations of NH_4^+-N were greater in soils fertilized with BBU, which promoted high NH_3 volatilization compared to soils fertilized urea (Jia et al. 2021). Nevertheless, the application of BBU significantly reduced N leaching and ultimately reduced total N loss of N fertilizer.

3.3 Benefits of BBFs for Crop Growth and Productivity

The application of BBFs can either increase or maintain crop productivity, depending on factors such as soil type, climate, feedstock variations, carbon content, and pyrolysis conditions during production (Álvarez Salas et al. 2024; Melo et al. 2024, 2022). A recent meta-analysis found that BBFs improved crop productivity by 10% at low application rates (an average of 0.9 t ha^{-1}) compared to fertilized controls (Melo et al. 2022). Furthermore, BBFs with high carbon concentrations, typically produced at elevated temperatures, and those applied to initially fertile soils resulted in even greater productivity.

One key advantage of BBFs is their gradual nutrient release, optimizing plant nutrition. For example, BBFs enriched in P and Mg significantly increased P uptake in grass shoots during the second and third cropping cycles compared to conventional phosphate fertilizers (TSP) (Lustosa Filho et al. 2020). In the second cycle, P uptake was 33.1 mg pot^{-1} with BBFs compared to only 15.4 mg pot^{-1} with TSP, demonstrating the slow release of P from BBFs, which was absent in conventional fertilizers. Schmidt et al. (2017) showed that adding BBFs synthesized with cow urine or NPK fertilizer significantly boosted yields compared to controls. Specifically, biochar enriched with NPK at low dosages (0.5 and 2 t ha^{-1}) resulted in a 20% average yield increase over NPK fertilizer alone. These improvements are attributed to biochar's role as a nutrient carrier, promoting slow nutrient release and minimizing losses.

BBF application enhances the availability of macro- and micronutrients in soil while reducing heavy metal accumulation in plants. For instance, rice husk biochar-coated fertilizers significantly improved nutrient and agronomic efficiency, increasing photosynthetic rates and yields while reducing cadmium accumulation in rice plants by over 60% (Tan et al. 2024). This reduction in toxic element uptake highlights the safety and efficacy of BBFs in crop production.

Economically, using BBFs offers noteworthy advantages. Studies indicate that reduced application rates can yield favorable economic and environmental benefits. For example, applying approximately 2,544 to 2,625 kg ha^{-1} of BBF, compared to traditional fertilization at 3,000 kg ha^{-1}, significantly improved yield and fertilizer efficiency in eggplant cultivation (Zhang et al. 2022). This demonstrates that BBFs effectively meet crop nutritional needs while enhancing overall farm profitability. Thus, transitioning to BBFs could be a crucial step toward achieving agricultural sustainability while addressing the growing global food demands.

4. The Role of Biochar in Enhancing Water Holding Capacity of Soils

One of the notable benefits of biochar is improving the water-holding capacity (WHC) of soils, which is crucial for agriculture, especially in water-scarce regions (Trancoso et al. 2024). This section examines how biochar affects WHC, the underlying mechanisms, and the challenges of large-scale application.

WHC refers to a soil's ability to retain water, essential for plant growth and ecosystem health (Bilias et al. 2024). WHC is influenced by soil structure, texture, organic matter, and suction potential. Common potentials used to evaluate WHC include 0, 6, 10, 33, 1000, and 1500 kPa. Various methods exist to assess WHC, primarily relying on gravimetric moisture determination after equilibration at specific suction potentials. For instance, WHC at 0 kPa measures total porosity, while values at 6, 10, and 33 kPa are linked to field capacity (FC), and 1500 kPa corresponds to the permanent wilting point (PWP). Teixeira and Behring (2017) outlined methodologies for measuring WHC in soils, while Teixeira et al. (2020) adapted these methods for biochar-amended soils. Other studies, including Blanco-Canqui (2017) and Omondi et al. (2016), reviewed the effects of biochar on total porosity and soil bulk density. Typically, adding biochar increases total porosity and decreases bulk density, although the extent of these effects varies with biochar feedstock, particle size, and soil type. Wang et al. (2017) found that biochar could increase or decrease infiltration rates, depending on soil characteristics, with sandy soils often experiencing a decrease. Conversely, Abrol et al. (2016) noted improved infiltration rates in crust-prone soils.

FC represents the soil moisture content after gravitational water has drained, while PWP indicates the moisture level at which plants cannot recover from wilting (Bilias et al. 2024). A global meta-analysis by Edeh et al. (2020) revealed that biochar generally increased available water content (AWC) by 28.5%, FC by 20.4%, and PWP by 16.7%. Razzaghi et al. (2020) showed that biochar significantly improved soil water content at FC in coarse-textured soils but had no effect in fine-textured soils, with plant-available water (PAW) improvements decreasing with soil fineness. In the field, actual soil moisture content is influenced not only by WHC but also by infiltration capacity. Gholamahmadi et al. (2023) found that biochar reduced runoff by 25%, particularly in temperate zones (28% reduction), while effects in (sub)tropical zones were not significant.

The mechanisms by which biochar affect WHC and PAW remain partially understood. A crucial factor is biochar's porosity, as its high internal surface area enables greater water retention (Basso et al. 2013). Sun and Lu (2014) suggested that biochar enhances soil aggregate stability, further improving water retention. Teixeira et al. (2017) demonstrated that biochar produced at higher pyrolysis temperatures holds more water at high potentials (≥1500 kPa) due to increased specific surface area and charge. Batista et al. (2018) found that enhanced cation exchange capacity and changes in zeta potential on biochar surfaces contribute to increased water retention through hydrated ion adsorption.

Biochar particle size significantly impacts WHC, with finer particles generally increasing gravimetric water content more efficiently than coarser ones (Verheijen et al. 2019; Blanco-Canqui, 2017; Glab et al. 2016). Liu et al. (2017) showed that smaller particle sizes alter pore space between particles and create more pores within the biochar, leading to higher WHC.

The pyrolysis temperature affects biochar's specific surface area (SSA), porosity, and hydrophobicity, influencing its water retention capacity. Teixeira et al. (2017) noted that higher pyrolysis temperatures typically enhance SSA and improve biochar's ability to hold water under high pressures. However, excessively high temperatures can clog micropores, reducing SSA (Brown et al. 2016). Chen et al. (2019) also found that low pyrolysis temperatures limit SSA due to retained volatiles and tars, which can fill biochar's internal pore structure.

The ideal pyrolysis temperature for producing biochar with high WHC depends on the desired porosity and water-holding characteristics for specific applications. Teixeira et al. (2021) summarized these findings, indicating the trade-offs involved in biochar production. While biochar shows potential for improving WHC, scaling its agricultural use presents challenges. Economic feasibility is a primary concern, with some studies reporting application rates as high as 195 Mg ha^{-1}, which would be cost-prohibitive for most farmers. The variability in biochar's effects on WHC underscores the importance of feedstock selection and pyrolysis methods. Tailoring biochar properties, such as pore size distribution and SSA, could enhance its effectiveness for specific soil types.

Not all soils respond equally to biochar applications; it tends to be more effective in coarse-textured soils than in fine-textured ones. Research suggests that tailor-made biochars produced by selecting specific feedstocks and adjusting pyrolysis conditions could optimize biochar's impact on water retention. The concept of "designer biochar" may help maximize soil-water storage for specific pressure ranges and soil types. Caution is needed when applying large quantities of biochar, especially from poorly studied feedstocks or unregulated pyrolysis techniques. The long-term effects of using non-certified biochars could adversely impact soil health (Verheijen et al. 2010, 2012; IBI, 2015; EBC, 2012).

Although still less explored, biochar-based fertilizers (BBFs) can be designed to enhance nutrient release while increasing soil water retention. Zheng et al. (2021) showed that BBFs improve peanut growth, yield, and stress tolerance under water-limited conditions. Applying 0.75 t ha^{-1} BBFs with moderate water stress increased yields by 22-28%, demonstrating their potential to enhance water and nutrient retention in arid regions. An et al. (2021) developed biochar-based slow-release P fertilizers via co-pyrolysis and co-polymerization, offering high water retention (94.2 g/g) and prolonged phosphorus release (75.8% efficiency after 60 days). These fertilizers improved plant growth, retained soil moisture, and are biodegradable, providing a sustainable solution for agriculture. Wen et al. (2017) studied biochar-based slow-release fertilizers (BSRFs) using microwave-assisted synthesis to enhance water retention and nutrient efficiency. BSRFs improved water-holding capacity (60.6%), reduced nitrogen leaching (10.3%), and provided controlled nitrogen release over 30 days. Pot experiments showed 64.2% nitrogen-use efficiency and improved crop growth. BSRFs are biodegradable and sustainably synthesized, offering significant potential for agriculture. Despite the benefits demonstrated under controlled conditions, fieldwork is needed to validate these findings and explore their economic feasibility.

5. Conclusions and Recommendations for Future Research

There is still a lack of field experiments exploring biochar application in conjunction with reduced chemical fertilizer use. However, studies indicate that a fertilizer reduction of up to 20% is generally safe when combined with biochar, and in some cases, even an increase in crop yield is observed. This suggests that biochar can enhance nutrient use efficiency. Long-term field studies are necessary to determine how long the effects of a single biochar application persist. Additionally, it is important to correlate the initial biochar quality and application rate to maximize fertilizer reduction while improving crop yields.

Biochar shows promise as a soil amendment, particularly for improving WHC in sandy and coarse-textured soils. However, its effectiveness can vary significantly based on feedstock, pyrolysis conditions, and soil type. While biochar's high porosity and SSA generally enhance soil WHC, some studies report minimal or negative effects, indicating a need for further research into the underlying mechanisms.

Future studies should focus on optimizing biochar and biochar-based fertilizers production, assessing its economic viability for large-scale agricultural applications, and tailoring its properties to enhance nutrient and water retention in specific soil types. Additionally, more research is needed to understand the long-term effects of biochar on environmental and soil hydrology, particularly across different climatic and management contexts. While biochar has potential for improving soil water management, especially in drought-prone areas, significant challenges remain to ensure its application is economically and environmentally sustainable. A properly designed biochar, applied at suitable rates, could provide a valuable solution for enhancing soil water retention, but this requires careful consideration of its properties and long-term impacts.

Acknowledgements

We acknowledge the financial support of the National Council for Scientific and Technological Development (CNPq, Brazil) for providing Research Scholarships to the authors Leônidas C.A. Melo, Wenceslau G. Teixeira, Edson M. Mattiello, and Cícero C. Figueiredo. We also acknowledge the financial support of CESAM by the Portuguese Foundation for Science and Technology FCT/MCTES (UID P/50017/2020+UIDB/50017/2020+LA/P/0094/2020), through national funds. We further acknowledge funding from the FCT (DOI: 10.54499/PTDC/EAM-AMB/0474/2020; DOI: 10.54499/PTDC/EAM-AMB/1509/2021) as well as author Frank G.A. Verheijen (2023.06689.CEECIND).

References

Abrol, V., Ben-Hur, M., Verheijen, F.G., Keizer, J.J., Martins, M.A., Tenaw, H. et al. 2016. Biochar effects on soil water infiltration and erosion under seal formation conditions: Rainfall simulation experiment. *J. Soils Sediments* 16(12): 2709–2719.

Almeida Leite, M.C., Ballotin, F.C., Lustosa Filho, J.F., Santos, W.O., Matias, P.C., Pogorzelski, D. et al. 2023. Activated $ZnCl_2$ biochar and humic acid as additives in monoammonium phosphate fertilizer: Physicochemical characterization and agronomic effectiveness. *Environ. Res.* 232: 115927.

Álvarez Salas, M., Sica, P., Rydgård, M., Sitzmann, T.J., Nyang'au, J.O., El Mahdi, J. et al. 2024. Current challenges on the widespread adoption of new bio-based fertilizers: Insights to move forward toward more circular food systems. *Front. Sustain. Food Syst.* 8: 1386680.

An, X., Wu, Z., Liu, X., Shi, W., Tian, F. and Yu, B. 2021. A new class of biochar-based slow-release phosphorus fertilizers with high water retention based on integrated co-pyrolysis and co-polymerization. *Chemosphere* 285.

Basso, A.S., Miguez, F.E., Laird, D.A., Horton, R., and Westgate, M. 2013. Assessing potential of biochar for increasing water-holding capacity of sandy soils. *GCB Bioenergy* 5: 132–143.

Batista, E.M.C.C., Shultz, J., Matos, T.T.S., Fornari, M.R., Ferreira, T.M., Szpoganicz, B. et al. 2018. Effect of surface and porosity of biochar on WHC aiming indirectly at preservation of the Amazon biome. *Nat. Sci. Rep.* 8: 10677–10686.

Bilias, F., Sewu, D.D., Woo, S.H., Anastopoulos, I., Verheijen, F., Lehmann, J. et al. 2024. Glossary of terms used in biochar research (IUPAC Technical Report). *Pure Appl. Chem.* 96(11): 1541–1572.

Blanco-Canqui, H. 2017. Biochar and soil physical properties. *Soil Sci. Soc. Am. J.* 81(4): 687–711.

Brown, R.A., Kercher, A.K., Nguyen, T.H., Nagle, D.C. and Ball, W.P. 2006. Production and characterization of synthetic wood chars for use as surrogates for natural sorbents. *Org. Geochem.* 37(3): 321–333.

Carneiro, J.S. da S., Ribeiro, I.C.A., Nardis, B.O., Barbosa, C.F., Lustosa Filho, J.F., and Melo, L.C.A. 2021. Long-term effect of biochar-based fertilizers application in tropical soil: Agronomic efficiency and phosphorus availability. *Sci. Total Environ.* 760: 143955.

Cayuela, M.L., van Zwieten, L., Singh, B.P., Jeffery, S., Roig, A., and Sánchez-Monedero, M.A. 2014. Biochar's role in mitigating soil nitrous oxide emissions: A review and meta-analysis. *Agric. Ecosyst. Environ.* 191: 5–16.

Chen, W., Meng, J., Han, X., Lan, Y. and Zhang, W. 2019. Past, present, and future of biochar. *Biochar* 1(1): 75-87.

Chew, J., Zhu, L., Nielsen, S., Graber, E., Mitchell, D.R.G., Horvat, J. et al. 2020. Biochar-based fertilizer: Supercharging root membrane potential and biomass yield of rice. *Sci. Total Environ.* 713: 136431.

Ding, Y., Liu, Y., Liu, S., Li, Z., Tan, X., Huang, X. et al. 2016. Biochar to improve soil fertility: A review. *Agron. Sustain. Dev.* 36(2).

Domingues, R.R., Sánchez-Monedero, M.A., Spokas, K.A., Melo, L.C.A., Trugilho, P.F., Valenciano, M.N. et al. 2020. Enhancing cation exchange capacity of weathered soils using biochar: Feedstock, pyrolysis conditions and addition rate. *Agronomy* 10(6): 1–17.

Dominguez, E.L., Uttran, A., Loh, S.K., Manero, M.-H., Upperton, R., Idris Tanimu, M. et al. 2020. Characterisation of industrially produced oil palm kernel shell biochar and its potential as slow release nitrogen-phosphate fertilizer and carbon sink. *Mater. Today Proc.* 31: 221–227.

Edeh, I.G., Mašek, O. and Buss, W. 2020. A meta-analysis on biochar's effects on soil water properties – New insights and future research challenges. *Sci. Total Environ.* 714: 136857.

Fachini, J., de Figueiredo, C.C. and do Vale, A.T. 2022. Assessing potassium release in natural silica sand from novel K-enriched sewage sludge biochar fertilizers. *J. Environ. Manage.* 314: 115080.

Fachini, J., de Figueiredo, C.C., do Vale, A.T., da Silva, J., and Zandonadi, D.B. 2024. Potassium-enriched biochar-based fertilizers for improved uptake in radish plants. *Nutr. Cycl. Agroecosyst.* 128: 415–427.

Gholamahmadi, B., Jeffery, S., Gonzalez-Pelayo, O., Prats, S.A., Bastos, A.C., Keizer, J.J. et al. 2023. Biochar impacts on runoff and soil erosion by water: A systematic global-scale meta-analysis. *Sci. Total Environ.* 871: 161860.

Głąb, T., Palmowska, J., Zaleski, T. and Gondek, K. 2016. Effect of biochar application on soil hydrological properties and physical quality of sandy soil. *Geoderma* 281: 11–20.

Grutzmacher, P., Puga, A.P., Bibar, M.P.S., Coscione, A.R., Packer, A.P., and de Andrade, C.A. 2018. Carbon stability and mitigation of fertilizer-induced N_2O emissions in soil amended with biochar. *Sci. Total Environ.* 625: 1459–1466.

Hossain, M.Z., Bahar, M.M., Sarkar, B., Donne, S.W., Ok, Y.S., Palansooriya, K.N. et al. 2020. Biochar and its importance on nutrient dynamics in soil and plant. *Biochar* 2(4): 379–420.

International Biochar Initiative (IBI). 2015. Standardized product definition and product testing Guidelines for Biochar That Is Used in Soil. http://www.biochar-international.org/sites/default/files/IBI_Biochar_Standards_V2.1_Final.pdf.

Jia, Y.M., Hu, Z.Y., Ba, Y.X., and Qi, W.F. 2021. Application of biochar-coated urea controlled loss of fertilizer nitrogen and increased nitrogen use efficiency. *Chem. Biol. Technol. Agric.* 8(3): 1–11.

Laird, D., Fleming, P., Wang, B., Horton, R. and Karlen, D. 2010. Biochar impact on nutrient leaching from a midwestern agricultural soil. *Geoderma* 158: 436–442.

Leite, A. do A., de Resende, I.A.D., Lago, B.C., Barrera, A.M.V., Montes, A.O., de Assis Pereira, T. et al. 2024. Increasing the fertilizer efficiency of potassium silicate by co-pyrolysis and chemical activation. *J. Soil Sci. Plant Nutr.* 24: 2249–2262.

Liu, Z., Dugan, B., Masiello, C.A. and Gonnermann, H.M. 2017. Biochar particle size, shape, and porosity act together to influence soil water properties. *PLOS One* 12(6): e0179079.

Lu, J., Li, Y., Cai, Y., Jiang, P., and Yu, B. 2023. Co-incorporation of hydrotalcite and starch into biochar-based fertilizers for the synthesis of slow-release fertilizers with improved water retention. *Biochar* 5: 1–13.

Luo, Z., Li, Y., Pei, X., Woon, K.S., Liu, M. et al. 2024. A potential slow-release fertilizer based on biogas residue biochar: Nutrient release patterns and synergistic mechanism for improving soil fertility. *Environ. Res.* 252: 119076.

Lustosa Filho, J.F., da S. Carneiro, J.S., Barbosa, C.F., de Lima, K.P., Leite, A. do A., and Melo, L.C.A. 2020. Aging of biochar-based fertilizers in soil: Effects on phosphorus pools and availability to Urochloa brizantha grass. *Sci. Total Environ.* 709: 136028.

Lustosa Filho, J.F., Penido, E.S., Castro, P.P., Silva, C.A. and Melo, L.C.A. 2017. Co-pyrolysis of poultry litter and phosphate and magnesium generates alternative slow-release fertilizer suitable for tropical soils. *ACS Sustain. Chem. Eng.* 5: 9043–9052.

Melo, L.C.A., Lehmann, J., da S. Carneiro, J.S. and Camps-Arbestain, M. 2022. Biochar-based fertilizer effects on crop productivity: A meta-analysis. *Plant Soil* 472: 45–58.

Melo, L.C.A., Silva, C.A., Sánchez-Monedero, M.Á., Jindo, K., Taherymoosavi, S., and Joseph, S. 2024. Biochar-based fertilizers, co-composting, and growing media. pp. 669–694. *In:* Johannes L. and S. Joseph (eds), *Biochar for Environmental Management, Science, Technology and Implementation*. Routledge, London.

Moradi, S., Rasouli-Sadaghiani, M.H., Sepehr, E., Khodaverdiloo, H., and Barin, M. 2019. Soil nutrients status affected by simple and enriched biochar application under salinity conditions. *Environ. Monit. Assess.* 191: 1-13.

Morais, E.G., and Silva, C.A. 2023. Novel slow-release NPK biochar-based fertilizers with acidulated apatite: Evaluation of the fertilization value in a short-term experiment. *J. Soil Sci. Plant Nutr.* 23: 4937–4954.

Ndoung, O.C.N., de Souza, L.R., Fachini, J., Leão, T.P., Sandri, D. and de Figueiredo, C.C. 2023. Dynamics of potassium released from sewage sludge biochar fertilizers in soil. *J. Environ. Manage.* 346: 119057.

Ning, C., Liu, R., Kuang, X., Chen, H., Tian, J., and Cai, K. 2022. Nitrogen fertilizer reduction combined with biochar application maintain the yield and nitrogen supply of rice but improve the nitrogen use efficiency. *Agronomy* 12(12).

Omondi, M.O., Xia, X., Nahayo, A., Liu, X., Korai, P.K. and Pan, G. 2016. Quantification of biochar effects on soil hydrological properties using meta-analysis of literature data. *Geoderma* 274: 28–34.

Qi, S., Yang, S., Lin, X., Hu, J., Jiang, Z., and Xu, Y. 2023. The long-term effectiveness of biochar in increasing phosphorus availability and reducing its release risk to the environment in water-saving irrigated paddy fields. *Agric. Water Manage.* 282.

Razzaghi, F., Obour, P.B., and Arthur, E. 2020. Does biochar improve soil water retention? A systematic review and meta-analysis. *Geoderma* 361: 114055.

Rizhiya, E.Y., Horák, J., Šimanský, V. and Buchkina, N.P. 2020. Nitrogen enriched biochar-compost mixture as a soil amendment to the Haplic Luvisol: Effect on greenhouse gas emission. *Biologia* (Bratisl.) 75: 873–884.

Schmidt, H., Pandit, B.H., Cornelissen, G., and Kammann, C.I. 2017. Biochar-based fertilization with liquid nutrient enrichment: 21 field trials covering 13 crop species in Nepal. *L. Degrad. Dev.* 28: 2324–2342.

Sim, D.H.H., Tan, I.A.W., Lim, L.L.P., and Hameed, B.H. 2021. Encapsulated biochar-based sustained release fertilizer for precision agriculture: A review. *J. Clean. Prod.* 303: 127018.

Sun, F., and Lu, S. 2014. Biochars improve aggregate stability, water retention, and pore-space properties of clayey soil. *J. Plant Nutr. Soil Sci.* 177(1), 26–33.

Tan, T., Xu, Y., Liao, X., Yi, Z., and Xue, S. 2024. Formulating new types of rice husk biochar-based fertilizers for the simultaneous slow-release of nutrients and immobilization of cadmium. *GCB Bioenergy* 16, e13142.

Teixeira, W.G., and Behring, S.B. 2017. Retenção de água no solo pelos métodos da mesa de tensão e da câmara de Richards. *In:* P.C. Teixeira, G.K. Donagemma, et al. (eds.), *Manual de métodos de análise de solo* (3rd ed.), pp. 34–46. Brasília: Embrapa.

Teixeira, W.G., Marques, J.D.O., and Verheijen, F. 2020. Water holding capacity of biochar and biochar-amended soils. pp. 61–83. *In:* J.J. Manyà & G. Gascó (eds.), *Biochar as a Renewable-based Material: With Applications in Agriculture, the Environment and Energy*. London: World Scientific Publishing Europe.

Teixeira, W.G., Marques, J.D.O., Steiner, C., and Flanagam, R. 2017. Water retention in bamboo and wood biochar produced at different temperatures. *In:* P.M. Drumond & G. Wiedman (Eds.), *Bambus no Brasil: Da biologia à tecnologia*, pp. 368–380. Rio de Janeiro: ICH – Embrapa.

Tian, X., Li, Z., Liu, Z., Wang, Y., Li, B., Zhang, K. et al. 2022. Combined effect of biochar and nitrogen fertilizer reduction on rapeseed productivity and nitrogen use efficiency. *Arch. Agron. Soil Sci.* 68(9): 1159–1174.

Trancoso, R., Syktus, J., Allan, R.P., Croke, J., Hoegh-Guldberg, O., and Chadwick, R. 2024. Significantly wetter or drier future conditions for one to two-thirds of the world's population. *Nat. Commun.* 15(1): 483.

Verheijen, F., Jeffery, S., Bastos, A.C., Van der Velde, M., and Diafas, I. 2010. *Biochar Application to Soils: A Critical Scientific Review of Effects on Soil Properties, Processes, and Functions*. EUR 24099, 162.

Verheijen, F.G., Montanarella, L., and Bastos, A.C. 2012. Sustentabilidade, certificação e regulamentação do biochar. *Pesq. Agropecu. Bras.* 47: 649–653.

Verheijen, F.G., Zhuravel, A., Silva, F.C., Amaro, A., Ben-Hur, M., and Keizer, J.J. 2019. The influence of biochar particle size and concentration on bulk density and maximum water holding capacity of sandy vs. sandy loam soil in a column experiment. *Geoderma* 347: 194–202.

Wang, L., O'Connor, D., Rinklebe, J., Ok, Y.S., Tsang, D.C.W., Shen, Z., and Hou, D. 2020. Biochar aging: Mechanisms, physicochemical changes, assessment, and implications for field applications. *Environ. Sci. Technol.* 54(23): 14797–14814.

Wang, T., Stewart, C.E., Ma, J., Zheng, J., and Zhang, X. 2017. Applicability of five models to simulate water infiltration into soil with added biochar. *J. Arid Land* 9(5): 701–711.

Wen, P., Wu, Z., Han, Y., Cravotto, G., Wang, J., and Ye, B.C. (2017). Microwave-assisted synthesis of a novel biochar-based slow-release nitrogen fertilizer with enhanced water-retention capacity. *ACS Sustain. Chem. Eng.* 5(8): 7374–7382.

Wu, P., Ata-Ul-Karim, S.T., Singh, B.P., Wang, H., Wu, T., Liu, C. et al. 2019. A scientometric review of biochar research in the past 20 years (1998–2018). *Biochar* 1(1): 23–43.

Zhang, M., Liu, Y., Wei, Q., Liu, L., Gu, X., and Gou, J. 2022. Biochar-based fertilizer enhances the production capacity and economic benefit of open-field eggplant in the Karst region of Southwest China. *Agriculture* 12: 1388.

Zhang, S., Zhou, J., Chen, J., Ge, T., Cai, Y., Yu, B. et al. 2023. Changes in soil CO_2 and N_2O emissions in response to urea and biochar-based urea in a subtropical Moso bamboo forest. *Soil Tillage Res.* 228: 105625.

Zhang, Z., Zhang, Z., Gong, Z., Li, T., Nie, T., Chen, P. et al. (2024). Biochar application combined with water-saving irrigation enhances rice root growth and nitrogen utilization in paddy fields. *Agronomy* 14(5).

Zheng, J., Wang, S., Wang, R., Chen, Y., Siddique, K.H.M., Xia, G., and Chi, D. 2021. Ameliorative roles of biochar-based fertilizer on morpho-physiological traits, nutrient uptake and yield in peanut (*Arachis hypogaea* L.) under water stress. *Agric. Water Manag*. 257.

Zhou, M., Ying, S., Chen, J., Peikun, J., and Yuanxin, T. 2021. Effects of biochar-based fertilizer on nitrogen use efficiency and nitrogen losses via leaching and ammonia volatilization from an open vegetable field. *Environ. Sci. Pollut. Res.* 28: 65188–65199.

Zhou, J., Tang, C., Kuzyakov, Y., Vancov, T., Fang, Y., Song, X. et al. 2024. Biochar-based urea increases soil methane uptake in a subtropical forest. *Geoderma* 449: 116994.

Zhou, Z., Gao, T., Van Zwieten, L., Zhu, Q., Yan, T., Xue, J. et al. 2019. Soil microbial community structure shifts induced by biochar and biochar-based fertilizer amendment to Karst calcareous soil. *Soil Sci. Soc. Am. J.* 83: 398–408.

CHAPTER

3

Biochar Adoption in Agriculture: From Ancestral Roots to Large Scale Opportunities

Deborah Aller[1*], Martha Lustosa Carvalho[2], Ines Serra Baucells[3] and Daniel Velasquez[4]

[1] Section of Soil and Crop Sciences, School of Integrative Plant Science, Cornell University, Ithaca, New York, USA
[2] Department of Soil Science, Luiz de Queiroz College of Agriculture, University of São Paulo, Piracicaba, SP 13418900, Brazil
[3] BIOSORRA Carbon Removal Inc., Nairobi, Kenya
[4] Fundación Senderos, Mérida, Venezuela

1. Introduction

The deliberate application of charred or pyrogenic organic materials has occurred for thousands of years as evidenced by anthropogenic dark earths found in both tropical and temperate environments globally (Asare 2022). These include the *Terra Preta de Indio* soils in central Amazonia (Lehmann et al. 2003; Glaser 2007), Sub-Saharan Africa (Solomon et al. 2016), and Europe (Nicosia et al. 2017). In all cases these soils are dark in color, found near human settlements, and exhibit physicochemical properties that differ from the surrounding soils.

Biochar is the modern term for this historic practice and was first coined by Lehmann and Joseph (2009) as 'charred organic matter applied to soil in a deliberate manner, with the intent to improve soil properties.' Biochar has become a technology of increasing interest for farmers globally. The literature on biochar use in agriculture is substantial, with numerous studies supporting the potential improvements biochar provides to soil health, crop yields, and carbon sequestration (Woolf et al. 2010; Joseph et al. 2021; Lehmann et al. 2021; Schmidt et al. 2021). In fact, over 30,000 scientific studies since 2010 have explored biochar technologies both in agriculture and beyond (Gelardi and Parikh 2021)

*Corresponding author: da352@cornell.edu

DOI: 10.1201/9781003467946-4

and a systematic review summarized the numerous positive impacts biochar has on agronomic parameters (Schmidt et al. 2021). These observed benefits of biochar for soil health and crop productivity are particularly true in regions of the world where soil degradation is high and sandy and acidic conditions dominate (Melo et al. 2022). Thus, biochar presents a significant tool for increasing the sustainability of agricultural systems, particularly in the context of smallholder farmers in developing countries who often have limited access to agriculture inputs including fertilizer.

The majority of small-holder farmers are in rural areas of developing nations. For rural economies the idea of bioeconomy or circular economy has been promoted by organizations such as the United Nations Food and Agriculture Organization (FAO) as one strategy to mitigate poverty. The FAO defines the bioeconomy as the "production, utilization, conservation, and regeneration of biological resources, including related knowledge, science, technology, and innovation, to provide sustainable solutions (information, products, processes and services) within and across all economic sectors and enable a transformation to a sustainable economy" (Global Bioeconomy Summit Communiqué, 2020). Pyrolysis and biochar production and use is one technology that could contribute positively to the bioeconomy of countries.

Considering the seemingly countless evidence that biochar positively impacts soils properties and crop productively as well as the possible implementation of biochar into the bioeconomy of the rural sector, the question remains – why has biochar production and use not been more widely adopted? What are the current barriers limiting greater adoption? What can be done to increase adoption?

In agriculture, extension has been a well-established field within the United States since the early 1900s as it is a core pillar of the land grant university mission. Globally, extension became more common in the later part of the 1900s because it was established as a core component of development programs and projects to create change particularly in rural areas (UNFAO 1997). While the term extension may have different interpretations, it has the goal of being community leader and community focused, it is meant to be dynamic and bring new science, technology, and best practices from institutions to the field. Extension's goal is to work with people and be a continuously evolving process. It is also meant to be a two-way street, a partnership and collaborative process between the extension agent providing technical assistance and farmers. Farmers have knowledge and experience that extension agents do not, and vice versa. Extension is critically important, particularly in rural areas where access to resources and training may be limited.

To this point, extension is a critical way to disseminate biochar production and use to rural economies. Unfortunately, there remains a substantial gap between our scientific understanding of biochar and real-world applications on farms of all sizes. A study in the U.S. suggests that 'bridging the gap between scientific research and the practical implementation of biochar-based practices in the field can hasten the adoption of biochar-based land practices' (Pierson et al. 2024).

While this insight is based on information gathered during a workshop with U.S. forestry land managers it holds true for agriculture and expanding biochar use worldwide. Increasing awareness and understanding of the practical applications of biochar as well as addressing the many barriers to adoption will expand biochar's use in the agricultural sector of developing countries.

2. Barriers to Adoption

Beyond the scientific community biochar appears as a single product. In reality, biochar encompasses many different products and the diversity of biochars, largely resulting from different feedstock and production conditions, leads to impacts that are not always straightforward. Thus, at a high level the variability of biochars and the complexity of biochar interactions with soils, crops, weather, and management practices has made it challenging to predict its impacts on crop yields (Dokoohaki et al. 2019) and limited its widespread adoption in agricultural systems (Seman-Varner et al. 2022). Agriculture has many inherent risks (e.g., weather, markets, pests) and it is important not to introduce further risks to farmers. Risk avoidance has been identified as a possible barrier to adoption of technologies like biochar particularly for smallholder farmers in developing countries who may already be living on the margin and prefer to remain conservative and stick to what they know (Munoz 2014).

One study in the Ecuadorian Amazon found that post-harvest processing of agricultural waste, in this case coffee husks, had not been considered as a resource for the local bioeconomy and certainly not as a resource that could enhance the socioeconomic status of farmers in the community (Heredia Salgado et al. 2021). A lack of knowledge and logistical constraints meant that agricultural waste was not seen as a potentially valuable product by small-holder farmers. Further, the cost of biochar production, which was required to meet certain criteria within this region was too high, meaning that biochar was unlikely to be used by farmers compared to manures and other lower cost soil amendments. Building pyrolysis facilities in rural communities was seen as a potential positive through the creation of jobs and improvement in the socioeconomic status of individuals. However, the limited technical capacity of individuals in rural communities was a challenge for establishing and maintaining pyrolysis facilities as training would be required. Although, Heredia Salgado and colleagues (2021) found that carbon markets could be a game changer, as the monetization of carbon sequestration services could offer a new income stream for farmers. Similarly in Peru, financially incentivizing the low-tech carbonization of organic wastes was found to be the quickest way for management operators in the developing world to adopt emission reduction technologies like biochar production (Aquije et al. 2022).

Several studies have been conducted in Sub-Saharan Africa that evaluated the determinants for whether a farmer uses or does not use biochar. Socio-economic factors including age, sex, education, income, occupation, beliefs, farm size, and feedstock availability were found to influence biochar implementation in

Tanzania and Zimbabwe (Gwenzi et al. 2015; Fridahl et al. 2021; Rogers et al. 2021). Specifically gender and traditional gender roles were a factor. In Tanzania, men were found to have greater access to resources, such as land, and other opportunities compared to women. This led men to apply biochar to perennial crops including coffee and bananas, while women applied biochar to annual crops such as maize and beans so they could provide food for their families (Rogers et al. 2021). Farmers over the age of 40 in Cameroon were found to be more willing to apply biochar than younger farmers due to greater access to resources including land, labor, and financing (Fru et al. 2018). In both the Sub-Saharan Africa context (Rogers et al. 2021) and the Latin America and Caribbean context (Munoz, 2014) it was found that beliefs and compatibility with local cultural values and traditions were important for the adoption of new technologies such as biochar. If a practice was found to be culturally incompatible, individuals or a group of people were less likely to adopt it. Further, biochar may be viewed as charcoal production which is deemed as something done by people of lower socioeconomic status.

Overall, a lack of formal education and poverty level were found to be key determinants in farmers' lack of utilizing any soil inputs including biochar in Haiti (Munoz 2014) and Tanzania (Rogers et al. 2021). Insufficient knowledge and awareness as well as the lack of experts to provide technical assistance limits the potential adoption of biochar production and use. Access to extension services and technical assistance in rural areas is critical for information dissemination and increasing adoption. The prospects for biochar to improve yields, minimize the impacts of drought, and stabilize variability within smallholder farmer production make it a great potential solution, but if farmers don't know about it, how to make it or use it, adoption will continue to be limited. For example, in Sub-Saharan Africa feedstocks are seen as more important for other purposes rather than creating biochar including cooking and animal feed (Rogers et al. 2021). However, through education and training farmers could be made aware of low-tech kilns that could be used for both cooking and biochar production (e.g., TLUD stoves). Additionally, land tenure, whether a farmer owns or rents the land affects how they use that land. In the case of smallholder farmers who often do not own the land it was found that they had no incentive to improve soil health including using technologies like biochar (Munoz 2014).

The lack of financial capital, limited access to credit, and the lack of low-cost long-term financing to support biochar production and use further contribute to low biochar adoption in developing countries (Gwenzi et al. 2015; Rogers et al. 2021). However, crop yield improvements were observed following biochar applications, mainly in lower-income tropical countries where applications were too higher-value crops using biochar produced from small-scale decentralized systems, meant that biochar use was financially feasible (Robb et al. 2019). Unfortunately, upfront investment to begin biochar production and enable application by farmers is often not there. Financial capacity related to farm size is also found to be a factor that limits or enables adoption. Larger farms have the space and financial capacity to try new technologies (Kibue 2018; Solidaridad 2022) but at the same time small farms require only a small amount of material input and application to a smaller amount of land takes less time (Rogers et al. 2021).

Furthermore, policy support is critical to the adoption and use of new technologies globally. The lack of enabling policies and legal frameworks has been found to limit the adoption of new technologies in Sub-Saharan Africa including biochar technology (Gwenzi et al. 2015). When governments are unaware of certain technologies such as biochar, no guidelines or clear policies are created to facilitate advancement of the new technology. The need for supportive policies and appropriate legal frameworks at the local and national levels are needed in developing countries and beyond to advance biochar technology.

Summary of Barriers to Biochar Adoption in Agriculture in Developing Countries

- Lack of education and awareness
- Cost
- Risk avoidance
- Beliefs
- Gender
- Age
- Land tenure
- Farm size
- Feedstock and biochar availability
- Insufficient technical assistance (initial and sustained)
- Lack of supportive policies and legal frameworks
- Availability of long-term financing
- Income and poverty level
- Occupation
- Lack of trust
- Perceptions and attitudes

3. Developing Country Examples

Country level examples and case studies are critical to validating the expected outcomes of biochar use in any environment. Given the substantial variability between farms, including differences in climate, biochar properties, edaphic properties, cropping systems, and management practices, there can be tremendous differences in biochar impacts once applied to soil. Showcasing real-world scenarios is important for increasing understanding and expanding the likelihood of biochar adoption by others.

3.1 Case Study – KENYA

Historical Use

The use of biochar in Kenya and East Africa has deep roots, intertwined with traditional agricultural practices aimed at enhancing soil fertility and crop productivity. Indigenous communities in Kenya have long utilized charcoal and ash as soil amendments, drawing on generational wisdom. These practices were primarily based on an intuitive understanding of how biochar could improve soil health, particularly in regions with nutrient-poor soils. In many parts of Kenya, particularly areas with high nutrient leaching and acidic soils, biochar was historically used to enhance soil structure, water retention, and nutrient availability (Kätterer et al. 2019).

While the systematic documentation of biochar's use has been limited, its incorporation into farming practices was common, especially as a short-term soil amendment for improving nutrient content. The benefits of biochar were largely intuitive, with communities understanding that ashes from burned organic matter could provide essential nutrients for the soil.

One of the earliest modern initiatives to introduce biochar in Kenya was led by re:char in 2008. This marked the beginning of a movement that saw further organizations and initiatives such as Biochar Life, Safi Organics, and collaborations with institutions like Cornell University, ICRAF (International Centre for Research in Agroforestry), and the Africa Biochar Partnership. These efforts created a framework for advancing biochar systems across Africa, supported by platforms like the International Biochar Initiative.

Pioneering Industrial-Scale Biochar Production in Kenya

In 2023, BIOSORRA emerged as a pioneer in the industrial-scale production of biochar in Kenya, becoming the first carbon removal company of its kind in the region. It has since become the largest industrial biochar company in East Africa, driving the modern adoption of biochar technology at scale (Carbon Herald 2023) and positioning itself as a key player in the growing carbon removal and regenerative agriculture sectors (MIT Solve 2022). With its innovative approach, BIOSORRA has built up 30 use cases and demonstration farms across Kenya. While only about 2% of Kenyan farmers are using biochar, these farms have been critical in showcasing biochar applications across different production environments. These on-the-ground demonstrations are instrumental in proving the viability of biochar for enhancing soil fertility, increasing crop yields, and improving resilience to drought (Figure 1).

BIOSORRA's product, Biochar Bora, has been proven to increase soil pH, improve cation exchange capacity (CEC), and enhance nutrient retention. These improvements are essential for increasing agricultural productivity, particularly in regions where nutrient availability is limited. Soil pH has shown to be a useful and easy to measure soil parameter to monitor in Kenya when deciding how much biochar to apply and for assessing performance. Improving soil pH in acidic soils common in Kenya and most of Sub-Saharan Africa is critical for improved agricultural productivity.

Figure 1. Example of farmer's impact and yield increase in coffee at a Kenyan demonstration farm.

Use Cases for Biochar

Soil Acidity Remediation: At a demonstration farm located in Nakuru, Kenya using Biochar Bora in maize production, soil pH levels increased from 4 to 7.5 in 10 months. In flower farms in Naivasha, which are cultivated in greenhouses with controlled conditions and farming practices, Biochar Bora did not have any significant effect after 12 months, likely due to the already fertile volcanic soils. On other farms and crops, including tomatoes, onions, horticulture, and tree plantations, an average increase in soil pH from 5.6 to 7.81 was observed after 8 months.

Increased Yields and Water Retention: Trials have demonstrated that the application of Biochar Bora can lead to significant increases in crop yields. For example, maize yields improved by 15-30% in fields treated with biochar compared to no-biochar control plots. Biochar is also highly effective in improving soil water retention, which is critical for regions where rain-fed agriculture dominates. Biochar's ability to retain water helps crops withstand periods of drought, reducing the vulnerability of farmers to climate change and weather extremes.

Farmers' economic Impact: Based on 30 BIOSORRA's case studies with farmers in Kenya, for each 1$ invested in Biochar Bora results in $0.66 in profit coming from cost savings that include reductions in fertilizer, lime, and water while ensuring higher quantity and quality of crops. The greatest yield increase at one farm was 150%, while the average yield increase across farms is 28%. This directly translates to farmers' families positively impacted and greater food security. In fact, farmer participants have said "biochar should be introduced to many other farmers" and "the plants were bigger and greener, the crops were healthier than without biochar and there was a high improvement in yield".

Employment and Community Benefits: Through its industrial-scale production and use of biochar, BIOSORRA has provided employment to hundreds of local community members. The company's approach fosters a circular economy, ensuring that local farmers and workers benefit from biochar's production and application. Notably, BIOSORRA is a female-led company, with an 80% female workforce, highlighting its commitment to gender equality and community development.

Other significant co-benefits of biochar include climate change mitigation and adaptation, by sequestering carbon in the soil, reduction of greenhouse gas emissions, economic empowerment, improved farmer livelihoods, water stewardship, food security, and circular bioeconomy.

Barriers to Biochar Adoption

Biochar adoption in Kenya is still in its early stages, but key efforts highlight its transformative potential. By addressing barriers through community engagement, product transparency, and scaling production, BIOSORRA is setting the stage for a future where biochar becomes a mainstream solution for sustainable agriculture and climate change mitigation in East Africa and beyond. The lessons learned in Kenya can provide valuable insights for biochar adoption elsewhere, paving the way for a global movement towards regenerative agriculture and carbon sequestration. Some of the key barriers to biochar production and use in Kenya include:

Perception of Biochar as a New Product: For farmers, biochar is a new and unfamiliar product. Biochar, with its various feedstocks, production methods, and site-specific applications, is complex and requires significant investment in increasing education and awareness. Farmers, especially those living on less than $2/day, are more focused on short-term solutions to agricultural resilience and may not be willing to invest in biochar without understanding its long-term benefits.

Variability in Biochar Quality: The quality of biochar varies depending on the feedstock used, the pyrolysis conditions, and the intended end-use. This inconsistency can make it difficult for farmers to trust the product and its efficacy in their specific agricultural contexts.

Scaling Challenges in Production: Biochar production is either artisanal or industrial. The artisanal method, which involves low-tech, small-scale production, is not viable for large-scale agricultural systems due to limited production capacity. On the other hand, industrial-scale production requires significant investment, and financial viability can be challenging without relying on carbon credits. Additionally, the carbon credit market, while offering potential revenue streams, is still evolving and often inaccessible to small-scale farmers and producers in Sub-Saharan Africa, while for the larger ones, the methods are still heavily focused on Global North solutions.

Opportunities for Expanding Biochar Adoption

To increase biochar adoption among farmers it is essential to recognize the paramount importance of understanding local community priorities and align products with specific land application needs. BIOSORRA for example, follows an engagement strategy that encompasses two key components: public involvement within the community and ensuring equitable labor and economic opportunities linked to the project.

Implementing the Participatory Informational Community Development (PICD) model, BIOSORRA empowers communities to discern the root causes of underdevelopment and collaboratively devise solutions. This approach establishes trust and provides valuable insights into local perspectives on issues and specific needs. Extensive on-site community observations, interviews, and public forums, including webinars, have facilitated a comprehensive understanding. A collaboration with the Minister of Agriculture of Muranga County since the beginning has enhanced community presence and credibility. Continuous efforts include in-person soil health events in local dialects, engagement with local NGOs, and visits by farmers from Ghana and Kenya and representatives from supplier partners.

Public engagement serves not only to establish community trust but also to garner continuous feedback from farmers and monitor crop growth. Regular engagement sessions, such as group discussions and town halls, provide a platform for ongoing dialogue. A proactive approach that includes regular check-ins with local community members and the presence of local employees to aid in language and cultural understanding is critical. Additionally, collaboration with local communities is fundamental for acquiring the necessary data to quantify the benefits of biochar compared to other practices. Transparency in presenting this data to the community and training local farmers broadens scientific knowledge around biochar, increasing access for future use in Sub-Saharan Africa.

3.2 Case Study – BRAZIL

Historical Use

In Brazil, the current research and use of biochar have a direct cultural connection to the *Terra Preta de Indio* soils, its ancient heritage, and modern re-discovery by scientists. The dark and fertile anthropogenic soils found in the Amazon Basin reflect the soil management practices of pre-Columbian Indigenous peoples. The region's high rainfall begets highly weathered soils which are nutrient-poor, acidic, and with naturally low CEC. The rapid decomposition of organic matter in the humid climate exacerbates low soil fertility, making the region far from ideal for crop cultivation. The original peoples solved these barriers to crop production through a "slash-and-char" process. This practice not only enriched the soil with nutrients but increased its CEC, improving nutrient retention and supply to crops. This agricultural system allowed the indigenous peoples to grow crops efficiently in the naturally nutrient-poor soils of the Amazon basin for generations.

The recognition of Terra Preta's anthropogenic origin significantly advanced the understanding of sustainable agriculture in tropical environments. Contemporary biochar research, informed by these ancient practices, has demonstrated how charcoal and organic matter improve soil structure, enhance nutrient retention, and promote long-term soil fertility. This rediscovery underscores the sophistication of indigenous agricultural techniques and offers valuable insights into addressing the challenges of soil degradation and nutrient depletion in modern tropical agriculture.

Feedstock Availability in Agriculture

Brazil shows substantial potential for biochar production due to its abundance of agricultural and forestry resources. Vast quantities of crop residues, including sugarcane bagasse and straw, corn stover, and rice husks, could serve as feedstock for biochar, as well as the woody waste materials (e.g., sawdust, woodchips, and branches) produced from the approximately 10 million hectares of eucalyptus and pine plantations. The repurposing of these residues as a soil amendment is particularly relevant given the country's efforts to restore degraded lands and improve soil health, especially under initiatives such as the Plano ABC (Low Carbon Agriculture Plan), which promotes sustainable practices like no-till farming and integrated crop-livestock-forestry systems.

In addition, manure from the livestock sector and agro-industrial residues from processing crops like coffee, citrus, and oil seeds further increase the availability of potential feedstocks for biochar. Brazil is also the world's largest producer of charcoal, and charcoal production generates fines, which, if repurposed for biochar, could reduce waste, mitigate environmental and social concerns, and offer a valuable product. Despite the potential for production and use, there is a significant lack of comprehensive data regarding the quantity of biochar currently produced in Brazil, its uses in agriculture, and the availability of information for farmers, hindering a full assessment of its contribution to sustainable land management and carbon sequestration efforts. Data from on-farm and small-scale production remains anecdotal, as official data is unavailable.

Coffee Cooperatives – Industrial and On-farm Examples

At the industrial scale, coffee is the main crop receiving biochar in Brazil. NetZero, a tech startup, operates two large-scale biochar production facilities in the country. Located in the state of Minas Gerais, the largest coffee producing region, the two facilities can produce a combined 8 Mg of biochar per year. The feedstock is coffee husks collected from farms associated with agricultural cooperatives that facilitate the sale of biochar back to farmers. The circular business model integrates the entire process from biomass sourcing, producing and distributing biochar, leading field trials, to working on research, development, and certification. This circular model of partnering with coffee cooperatives works in areas with a large density of organized farmers who are geographically close to industrial infrastructure could work in other coffee-producing regions including São Paulo, Espírito Santo, and Bahia.

Coffee is a labor-intensive perennial agricultural crop with high fertilization rates, and biochar can reduce the need for industrial fertilizers. Biochar's soil enhancement properties can help ensure higher resilience of coffee plants to the abiotic stressors becoming more frequent in Brazil due to climate change, such as droughts and erratic rainfall. The use of coffee husks for biochar production is an economic and environmental positive for farmers with no agronomic disadvantage, as opposed to straw and stover removal in other crops that are operationally costly and can lead to soil degradation when mismanaged. The positive impact on soil health can increase coffee yield as well, as reported by NetZero's partner farmers. An unpublished study by NetZero and researchers at the Federal University of Lavras in Minas Gerais found that using biochar improved plant vegetative development (12.6% larger) and increased productivity by an average of 20%.

Not all farmers, however, are open to off-farm biochar solutions. For instance, some farmers use husks as fuel for coffee dryers, and off-farm biochar production would incur additional costs. However, on-farm production is hindered by equipment, as a large portion of coffee husks produced are completely burned to ash. However, Palinialves, a Brazilian agricultural machinery company, recently unveiled an on-farm pyrolysis unit that can cycle the heat produced in the oven to the dryers. The oven has an output of 1000-1200 L of biochar per day and is portable so can be moved around the farm, optimizing machinery use and maximizing efficiency in processing waste generated during harvest.

Additionally, coffee is often grown on steep terrain, with topography limiting the return of husks back to the crop. Pyrolyzing husks significantly reduces the volume of residue, making transportation to these areas more feasible. While application in steep areas remains manual, farmers in flatter areas can use solid spreaders, commonly employed for limestone application, to distribute biochar.

Partnerships for Biochar Introduction in Smallholder Agriculture

North: The status of the açai berry as a "superfood" with its introduction into North American and European markets drove production in Brazil's northern region from 145,800 tons in 1987 to 1.9 million tons in 2022 (FAPESPA, 2024). The state of Pará is currently the main producer of the berry, harvested by smallholders and communities who live alongside rivers in the Amazon. The increase has led to significant environmental challenges, particularly due to the improper disposal of açaí seeds. These seeds are often discarded in public areas or landfills, contributing to the pollution and sedimentation of water bodies. To address this issue, agronomist Moisés Mendonça developed an affordable kiln prototype that converts açaí seeds into biochar. The kiln is constructed using inexpensive and widely available materials, such as repurposed 200 L and 100 L drums, and uses combustible materials like pruned branches as fuel (de Souza Mendonça et al. 2024). The kiln design enables local farmers to produce biochar independently for their own use and for commercialization. Mendonça has been working to improve the kiln design in partnership with local producers in the city of Castanhal, Pará.

South: The Agricultural Research and Rural Extension Company of the State of Santa Catarina (Epagri) has partnered with vegetable producers in the state to introduce biochar into no-till onion production. The state has several charcoal plants that usually discard the smaller particles with no commercial use. Wood biochar, which in most cases is available free of charge, is spread on top of the crop residue and has been shown to increase onion yield (Higashikawa et al. 2023). According to Higashikawa, farmers from Santa Catarina are also applying biochar to other vegetable crops including common beans.

Current Barriers to Adoption

One of the primary barriers to biochar adoption in Brazil is the limited supply and market availability. Until this constraint is addressed, widespread implementation remains challenging and limited to specific regions and crops. As supply increases with investments such as the new facilities built by NetZero, cost will likely become the next major obstacle. Biochar represents a medium- to long-term investment in soil health, yet many farmers still prioritize land acquisition over soil amendments when making financial decisions. Another significant challenge is the scarcity of concrete data on biochar's agronomic benefits under specific soil and climate conditions in Brazil. Demonstrating its effectiveness for key crops is essential to justify its adoption by farmers. Finally, most professionals in the agricultural sector are not trained in soil health and are skeptical of innovation in the field, further hindering adoption. For instance, while the Brazilian forestry industry is highly innovative in industrial processes, its investment cycles often extend over decades. Consequently, biochar is not considered a priority for the sector. This is particularly unfortunate since forestry residues are among the most promising and cost-effective feedstocks for biochar production in Brazil.

3.3 Case Study – VENEZUELA

Fundación Senderos, a local NGO based in La Mucuy, Mérida, Venezuela, is leading a transformative biochar initiative. Initially supported by the GEF Small Grants Programme and national organizations united under Impact Hub Caracas, the foundation employs artisanal flame-cap technologies, such as the Kon Tiki and RoCC kilns, to convert biomass waste—coconut shells, maize cobs, coffee branches, forest material, and bamboo—into biochar-based fertilizers (BBF).

In collaboration with Woodgas International, Fundación Senderos also delivers Fabstoves, a clean, biochar-producing cookstove that benefits both farmers and households in the community. The clean cooking solution contributes to reducing emissions while providing a sustainable energy source. The foundation's headquarters, Casa La Lumbre, reflects a deep commitment to sustainability by integrating eco-tourism, permaculture, and circular economy practices, serving as both a model of sustainable living and a regional innovation hub. Additionally, Fundación Senderos leads educational workshops at local schools, educating children and youth about the importance of soil health and the benefits of biochar.

Biochar in Coffee Production and Its Impact

Farmers in La Mucuy, organized through the Mucu-y-Café Association and supported by Fundación Senderos, are integrating biochar at various stages of coffee cultivation (Figure 2). The application of biochar in nurseries enhances soil health and fosters better biological management practices. By utilizing low-tech kilns like Kon Tiki, farmers can produce biochar on-site from local agricultural waste, converting it into BBF. This approach reduces the dependency on external agricultural inputs, supports circular agricultural practices, and enhances soil fertility, which is critical to the long-term sustainability of coffee farming. The Mucu-y-Café Association plans to further expand its specialty coffee production, building on the success of biochar integration.

Scaling Production and Application

The scale of biochar production in La Mucuy is currently small to medium, with significant potential for growth. The biochar produced is being applied across approximately 10,000 coffee trees, illustrating its broad use and positive impact. At Hacienda La Secreta, more than 7,000 coffee seedlings were propagated in 2024 using different ratios of biochar-enriched soils, further demonstrating the scalability of the initiative. As the community adopts biochar more widely,

Figure 2. Biochar production and use in La Mucuy, Venezuela.

the region is poised to increase both the scale and impact of its biochar-based agricultural solutions.

Bridging Barriers: Digital Measurement, Reporting, and Verification (dMRV) Role in Carbon Markets and Scaling Biochar Adoption in Venezuela

The adoption of biochar in Venezuela faces significant barriers, particularly related to the limited demand for the product due to its scarce availability at the industrial scale needed to meet market requirements. This creates a cycle where demand cannot grow without a reliable supply, and large-scale production is hampered by insufficient demand. Agrocognitive, a Venezuelan startup, is addressing this challenge by playing a key role in building a comprehensive ecosystem that connects stakeholders across the biochar supply chain. Through its Decarbonization Lab for the Coffee and Cocoa Clusters, they set out to demonstrate the economic and environmental benefits of biochar for these sectors while leveraging its dMRV platform to ensure traceability and attract investment.

By establishing collaborations with local producers, research institutions, international organizations, and policymakers, Agrocognitive is creating the conditions necessary for scaling up biochar production and fostering demand. Its efforts focus on integrating biochar into the coffee and cocoa sectors in Mérida as a niche market, showcasing its potential to enhance soil health, increase yields, and reduce chemical inputs. This targeted approach not only addresses immediate barriers but also lays the groundwork for broader adoption, positioning Venezuela to tap into emerging opportunities in carbon markets and sustainable agriculture.

4. Increasing Adoption in Developing Countries

While there are clear barriers to adoption of biochar technology in developing countries there also exist numerous opportunities to decrease those barriers and increase adoption. Increasing awareness of what biochar is, how it is made, how it can be used, and its potential positive impacts through greater education, training, and outreach to farmers, policy makers, extension agents, and other stakeholders critical to agriculture would be a step forward in supporting biochar production and use. Once policy makers are aware of technology it gives them the opportunity to create enabling policies, guidelines, and legal frameworks to advance the use of technology if deemed appropriate. But without awareness there is no chance of change, even though this change may take years. The training of both public and private technical service providers (extension agents, crop consultants, etc.) would mean knowledge about biochar could be accessible to both large and small farms. While farmers who can afford to pay for private services are at an advantage in terms of access to assistance, the accessibility of public assistance may change as governments change and when considering cell phone and internet availability to be able to reach rural communities that could not be reached before.

5. Role of Agriculture Extension

Agriculture extension provides non-formal education and learning activities to farmers and other agricultural stakeholders worldwide. Extension is critical for bringing essential, practical research-based information to farmers. The core of extension is openness, accessibility, and service. Extension agents bring evidence-based science and new technologies to farmers and offer expertise and resources to address community needs. Technology transfer to farmers is needed for sustainable development in rural communities. With biochar for example, it is standard practice for most smallholder farmers worldwide to burn agricultural residues and organic materials, but through demonstrations and trainings a change in practice and mindset could shift burning agricultural wastes to charring them and applying them back to the soil for improved production.

Unfortunately, while extension exists in most countries, rural extension and technical assistance is often insufficient. In Brazil, smallholder farmers are unable to obtain technical assistance because they rely on public service while large landowners can pay for private technical assistance. The level of understanding regarding the importance and value of technical assistance also plays a role in who is willing to invest in private technical assistance versus relying on inadequate public services. Specifically, government agents predominantly provide public technical assistance in rural areas but between 2006 and 2017 the services provided to farmers decreased from 43% to 38%, while those farmers who could afford to pay for private technical assistance increased to 31% in 2017 (Solidaridad, 2022). In the Brazilian State of Pará it was found that 95% of smallholder farmers had no access to rural extension or technical assistance services (Solidaridad, 2022).

5.1 Path Forward

Empowering farmers and community members to produce and use biochar through hands-on training and demonstration projects helps foster sustained use of the technology. Farmers must trust who they receive information from and feel comfortable about using a technology for them to begin and continue using it. Farmer cooperatives can play an important role in supporting farmers in their biochar journey and can help contribute to price stabilization of agricultural products especially as it relates to large private companies (Heras-Saizarbitoria and Basterretxea, 2016). Furthermore, communicating to farmers at the appropriate level is essential. Practical information about biochar needs to be provided in a language that end users and technical service providers understand. Technical jargon limits the accessibility of information, especially considering the lack of formal education of many smallholder farmers in developing countries. Farmers want to see how a product works for themselves and have answers to practical questions such as how much do I apply? How do I apply it? When do

I apply the biochar? Where do I apply it? In fact, Fridahl and colleagues (2021) found in Tanzania that over the long-term education, awareness, and persistent demonstration were key to the sustained implementation of biochar technology. Having local field trials and demonstrations of biochar production and use increases confidence of farmers, and local availability of biochar motivated smallholder farmers to use it on their fields in Tanzania (Rogers et al. 2021). Farmers want to see biochar applications to their soils and crops, using their management practices, and under local weather conditions.

The scale of production of biochar can play a big role in increasing adoption in developing countries. Considering financing limitations and the lack of public investment in large-scale systems in places like Sub-Saharan Africa (Gwenzi et al. 2015) it makes more sense to focus on small-scale production. Small-scale pyrolysis technologies have a cost advantage over medium and large-scale facilities as well because they are generally closed systems and farmers can produce biochar from their own or local wastes and apply it back to fields (Robb et al. 2019). Small scale on-farm production of biochar reduces dependence on outside producers, facilitates organic wastes management, minimizes costs, and recycles carbon and nutrients directly on farm. Additionally, unlike in temperate environments where crop yield improvements are more limited, the positive effects of biochar on crop yields are substantially greater in tropical soils (Crane-Droesch et al. 2013; Jeffery et al. 2017), which can offset the financial costs further (Robb et al. 2019).

Another relevant consideration for smallholder farmers in developing countries is improved air quality for better respiratory health. Cooking indoors with limited ventilation using charcoal or other solid fuel sources is still a common practice in the developing world and a major health concern (Bonjour et al. 2013). Small kilns can double as cook-stoves and drastically reduce the smoke produced and inhaled by household members. In Kenya, one study found that using gasifier cookstoves that simultaneously produced biochar improved energy use efficiency and indoor air quality. Specifically, the levels of carbon monoxide, carbon dioxide, and fine particulate matter, were overall reduced compared to the standard cooking method by 57%, 41%, and 79%, respectively (Gitau et al. 2019). The more widespread use of biochar producing cookstoves in rural communities could have huge positive implications for public health. But individuals and government officials need to be aware of this technology as an option for its production and use to be expanded.

6. Conclusions

While in theory biochar is not a new technology it is a new technology for most individuals worldwide. Clear barriers to adoption exist among rural communities in developing countries but a path forward exists to increase adoption. Biochar is an important tool in the push towards more climate-smart agriculture. The evolution of carbon markets in the coming years may play a large role in expanding biochar production and use. But first, the economic, social, and political barriers to biochar production and use need to be lowered for biochar technology to have a chance at replacing current technologies. For this to occur, individuals need to be aware of biochar, thus education and awareness must be greatly increased. Continuing to improve communication pathways between farmers, NGOs, government agencies, private consultants and public institutions can help foster more technical assistance and rural extension.

In areas where access to other solutions is limited biochar technology has great potential. Where smallholder farmers do not have the financial means or access to fertilizer, biochar offers a promising solution especially when produced locally providing an additional waste-management tool for farmers. For all family members replacing open indoor cooking fires with small cookstoves that double as biochar production kilns can have huge public health benefits. Biochar has tremendous potential to improve livelihoods in developing countries and contribute to a more circular bioeconomy.

References

Asare, M.O. 2022. Anthropogenic dark earth: Evolution, distribution, physical, and chemical properties. *European Journal of Soil Science* 73(5): e13308. https://doi.org/10.1111/ejss.13308

Aquije, C., Schmidt, H.-P., Draper, K., Joseph, S., and Ladd, B. 2022. Low tech biochar production could be a highly effective nature-based solution for climate change mitigation in the developing world. *Plant and Soil* 479(1–2), 77–83. https://doi.org/10.1007/s11104-021-05159-6

Bonjour, S., Adair-Rohani, H., Wolf, J., Bruce, N.G., Mehta, S., Prüss-Ustün et al. 2013. Solid fuel use for household cooking: Country and regional estimates for 1980–2010. *Environmental Health Perspectives* 121: 784–790.

Carbon Herald, 2023. BIOSORRA Launches the Biggest Biochar-Producing Plant in East Africa. Accessed on 12/01/2024. https://carbonherald.com/biosorra-launched-the-biggest-biochar-producing-plant-in-east-africa/

Crane-Droesch, A., Abiven, S., Jeffery, S., and Torn, M.S. 2013. Heterogeneous global crop yield response to biochar: A meta-regression analysis. *Environmental Research Letters* 8(4): 044049. https://doi.org/10.1088/1748-9326/8/4/044049

de Souza Mendonça, M., de Melo, W.J., de Melo, G.M.P., Bertipaglia, L.M.A., Araujo, A.S.F., Reis, I.M.S. et al. 2024. Açai seed biochar improves soil quality and black pepper seedling development in the Amazon region. *Journal of Environmental Management*, 367, 121752. https://doi.org/10.1016/j.jenvman.2024.121752

Dokoohaki, H., Miguez, F.E., Laird, D., and Dumortier, J. 2019. Where should we apply biochar? *Environmental Research Letters* 14(4), 044005. https://doi.org/10.1088/1748-9326/aafcf0

FAPESPA. 2024. Conjuntura Econômica do Açaí Paraense (Technical Report 1; p. 23). https://www.fapespa.pa.gov.br/wp-content/uploads/2024/06/Nota-Tecnica-Conjuntura-da-Economia-do-Acai-2024.pdf

Food and Agriculture Organization of the United Nations (UNFAO). 1997. Guide to Extension Training. Chapter 2. *In:* Oakley, P., and Garforth, C. (eds.), *Understanding Extension.* https://www.fao.org/4/t0060e/T0060E03.htm#2.%20Understanding%20 extension

Fridahl, M., Haikola, S., Rogers, P.M., and Hansson, A. 2021. Biochar deployment drivers and barriers in least developed countries. *In:* Leal Filho, W., Luetz, J., Ayal, D. (eds.), *Handbook of Climate Change Management: Research, Leadership, Transformation.* Springer, pp. 1–30.

Fru, B.S., Angwafo, T.E., Martin, T.N., Francis, N.A., and Ngome, T.P. 2018. Environmental and socio-economic feasibility of biochar application for cassava production in the bimodal rainforest zone of Cameroon. *International Journal of Rural Development Environment and Health Research* 2: 1–9.

Gelardi, D.L., and Parikh, S.J. 2021. Soils and beyond: Optimizing sustainability opportunities for biochar. *Sustainability* 13: 10079. https://doi.org/10.3390/su131810079

Gitau, J.K., Sundberg, C., Mendum, R., Mutune, J., and Njenga, M. 2019. Use of biochar-producing gasifier cookstove improves energy use efficiency and indoor air quality in rural households. *Energies* 12(22): 4285. https://doi.org/10.3390/en12224285

Glaser, B. 2007. Prehistorically modified soils of central Amazonia: A model for sustainable agriculture in the 21st century. *Philosophical Transactions of the Royal Society, B: Biological Sciences*, 362: 187–196.

Global Bioeconomy Summit Communiqué, 2020. Expanding the Sustainable Bioeconomy Vision and Way Forward. Communiqué of the Global Bioeconomy Summit 2020.

Gwenzi, W., Chaukura, N., Mukome, F.N.D., Machado, S., and Nyamasoka, B. 2015. Biochar production and applications in Sub-Saharan Africa: Opportunities, constraints, risks and uncertainties. *Journal of Environmental Management* 150: 250–261. https://doi.org/10.1016/j.jenvman.2014.11.027

Heras-Saizarbitoria, I., and Basterretxea, I. 2016. Do co-ops speak the managerial lingua franca? An analysis of the managerial discourse of Mondragon cooperatives. *Journal of Cooperative Organization and Management* 4: 13–21.

Heredia Salgado, M.A., Säumel, I., Cianferoni, A., and Tarelho, L.A.C. 2021. Potential for farmers' cooperatives to convert coffee husks into biochar and promote the bioeconomy in the north Ecuadorian Amazon. *Applied Sciences* 11(11), 4747. https://doi.org/10.3390/app11114747

Higashikawa, F.S., Silva, C.A., Carducci, C.E., Jindo, K., Kurtz, C., de Araújo, E.R. et al. 2023. Effects of the application of biochar on soil fertility status, and nutrition and yield of onion grown in a no-tillage system. *Archives of Agronomy and Soil Science* 69(2): 212–227. https://doi.org/10.1080/03650340.2021.1978073

Jeffery, S., Abalos, D., Prodana, M., Bastos, A.C., Van Groenigen, J.W., Hungate et al. 2017. Biochar boosts tropical but not temperate crop yields. *Environmental Research Letters* 12(5): 053001. https://doi.org/10.1088/1748-9326/aa67bd

Joseph, S., Cowie, A.L., Van Zwieten, L., Bolan, N., Budai, A., Buss, W. et al. 2021. How biochar works, and when it doesn't: A review of mechanisms controlling soil and plant responses to biochar. *GCB Bioenergy* 13(11): 1731–1764. https://doi.org/10.1111/gcbb.12885

Kätterer, T., Roobroeck, D., Andrén, O., Kimutai, G., Karltun, E., Kirchmann, H. et al. 2019. Biochar addition persistently increased soil fertility and yields in maize-soybean rotations over 10 years in sub-humid regions of Kenya. *Field Crops Research* 235: 18–26. https://doi.org/10.1016/j.fcr.2019.02.015

Kibue, G.W. 2018. Use of biochar for increased crop yields and reduced climate change impacts from agricultural ecosystems: Chinese farmers' perception and adoption strategy. *African Journal of Agricultural Research* 13: 1063–1070.

Lehmann, J., Kern, D.C., Glaser, B., and Woods, W.I. 2003. *Amazonian Dark Earths: Origin, Properties, Management*. Kluwer Academic Publishers.

Lehmann, J., Cowie, A., Masiello, C.A., Kammann, C., Woolf, D., Amonette et al. 2021. Biochar in climate change mitigation. *Nature Geoscience* 14(12): 883–892. https://doi.org/10.1038/s41561-021-00852-8

Melo, L.C.A., Lehmann, J., Carneiro, J.S., and Camps-Arbestain, M. 2022. Biochar-based fertilizer effects on crop productivity: A meta-analysis. *Plant Soil* 472: 45–58. https://doi.org/10.1007/s11104-021-05276-2

Munoz, L. 2014. *Spreading the Char: The Importance of Local Compatibility in the Diffusion of Biochar Systems to the Smallholder Agriculture Community Context*. Pomona College.

Nicosia, C., Devos, Y., and Macphail, R.I. 2017. European dark Earth. *In:* C. Nicosia & G. Stoops (eds.), *Archaeological Soil and Sediment Micromorphology* (1st ed., pp. 331–343). Wiley. https://doi.org/10.1002/9781118941065.ch32

Pierson, D., Anderson, N., Brewen, J., Clark, N., Hardy, M.C., McCollum et al. 2024. Beyond the basics: A perspective on barriers and opportunities for scaling up biochar production from forest slash. *Biochar*, 6(1), 1. https://doi.org/10.1007/s42773-023-00290-2

Robb, S., Joseph, S., Abdul Aziz, A., Dargusch, P., and Tisdell, C. 2020. Biochar's cost constraints are overcome in small-scale farming on tropical soils in lower-income countries. *Land Degradation & Development*, 31(13): 1713–1726. https://doi.org/10.1002/ldr.3541

Rogers, P.M., Fridahl, M., Yanda, P., Hansson, A., Pauline, N., and Haikola, S. 2021. Socio-economic determinants for biochar deployment in the Southern Highlands of Tanzania. *Energies* 15(1): 144. https://doi.org/10.3390/en15010144

Schmidt, H.P., Kammann, C., Hagemann, N., Leifeld, J., Bucheli, T.D., Sánchez Monedero, M.A. et al. 2021. Biochar in agriculture – A systematic review of 26 global meta-analyses. *GCB Bioenergy* 13(11): 1708–1730.

Seman-Varner, R., Hassebrook, C., Zilberman, D., Brown, R., Paul, B., Winstel, L. et al. 2022. *Scaling Sustainable Biochar Research & Commercialization for Agriculture & Conservation Benefits: A Summary from a Stakeholder Convening* [White paper]. Washington, DC: AFT-FFAR-NCAT.

Solidaridad. 2022. *The Technical Assistance and Rural Extension (TARE) Void: Pathways for the Socioeconomic and Environmental Inclusion of Smallholder Farming*. São Paulo: Solidaridad, 2022.

Solomon, D., Lehmann, J., Fraser, J., Leach, M., Amanor, K., Frausin, V. et al. 2016. Indigenous African soil enrichment as a climate-smart sustainable agriculture alternative. *Frontiers in Ecology and the Environment*, 14: 71–76.

Woolf, D., Amonette, J.E., Street-Perrott, F.A., Lehmann, J., and Joseph, S. 2010. Sustainable biochar to mitigate global climate change. *Nature Communications*, 1(1), 56. https://doi.org/10.1038/ncomms1053

Section 2 – Biochar: Unlocking Applications and Carbon Credit Pathways

Biochar has converted into a truly multifunctional product, moving far beyond agriculture to become part of a carbon cascade with applications across diverse industries. Each step in this cascade creates new opportunities for innovation and impact. The careful selection of feedstocks and understanding of biochar's behavior in soil are crucial to crafting a product that not only qualifies for carbon credits but also delivers real value in its end uses. For those pursuing biochar carbon credits, knowledge is power—it can save time, money, and unnecessary setbacks. Success in this field requires patience, rigor, and often the guidance of experienced experts, but the rewards—both for business and the planet—are worth the effort.

KEY LESSONS

- In an era of cascading crises caused by accelerating climate change, biochar provides cascading opportunities for enterprise and innovation.
- Biochar's unique stability turns soils into enduring carbon sinks, delivering climate benefits that are far more durable than those of conventional amendments.
- Successful biochar carbon credits depend on balancing patience, rigor, and trust-building across every stage.

CHAPTER

4

Carbon Cascades: Biochar Raw Material and Its Multiple Applications

Albert Bates[1]* and Kathleen Draper[2]

[1] Global Village Institute for Appropriate Technology, PO Box 90, Summertown, TN 38483

[2] Ithaka Institute for Carbon Intelligence, Canandaigua, NY 14564

1. Introduction

In its 2022 *Special Report on 1.5 Degrees*, the United Nations Intergovernmental Panel on Climate Change (IPCC) explained that climate warming from anthropogenic emissions from the pre-industrial period to the present will persist for centuries to millennia despite best efforts at emissions reductions this century and beyond. Excess greenhouse gases' atmospheric and oceanic persistence will augur ongoing impacts including sea level rise, extreme weather, pressure on global food supply, loss of ecosystems and biodiversity, and mass migration. The report called for the "upscaling and acceleration of far-reaching, multilevel and cross-sectoral climate mitigation" (Masson-Delmotte et al. 2019).

Biochar will be a critical carbon removal strategy, not only because it offers co-benefits that can accelerate adoption, such as in regenerative agricultural systems, but also because it can produce many other valuable products that are part of a circular carbon economy advancing economic and social development in the coming century. Transitioning from an extractive, carbon-emitting global economy to a 21st-century circular carbon economy will require a well-thought-out, whole-systems design approach to the human-built environment.

An essential concept in the carbon circular economy is the idea of "cradle to cradle"—that the end-of-life state for one product should transition into the

*Corresponding author: biochar@thefarm.org

DOI: 10.1201/9781003467946-6

birth of another. There is no such thing as "waste". This approach is helpful for climate-neutral or net-sequestration urban planning, sustainable water resource management, and regenerative farming and forestry. A circular carbon economy must link biochar process steps in ways that are locally sensitive to available biomasses, transportation distances, optimal plant design, and value chains to contribute to a regional bioeconomy.

A second essential element is "exergy"—a concept from thermodynamics that relates to the maximum practical work that can be extracted from a system as it reaches equilibrium with its surroundings. By using exergy analysis, urban planners can design more efficient buildings, district heating/cooling systems, and energy distribution networks; businesses can capture and utilize low-grade waste heat from industrial processes or power generation, integrating that into district heating systems or other valuable applications; and water managers can optimize water treatment, distribution, and reuse, reducing energy consumption and net system losses. Exergy concepts can inform strategies for managing urban heat islands and improving thermal comfort in outdoor spaces, such as strategically placing green spaces or water features. Exergy principles can inform the design of more efficient transportation networks, including integrating electric vehicles and optimizing public transit systems. In the context of biochar, exergy analysis supports a circular carbon model by helping identify opportunities for cascading energy and material use, eliminating waste, and valorizing climate mitigation.

1.1 Carbon Cascades

A carbon atom in the excited state has four valence electrons. After breaking apart under high temperature or pressure, carbon atoms may rearrange to form polymer chains, lattices, matrices, rings, or tetrahedra. Applying pressure to graphite, for example, can cause it to transform into a diamond with a more compact structure. Bonded together to form a cubic lattice of 3D tetrahedra, it minimizes the system's overall energy. The most stable form of carbon under given conditions is the one with the lowest energy requirement. In graphite, the carbon atoms are arranged into a honeycomb lattice with an average covalent bond length of 0.14 nm in numerous graphene layers. The interplanar spacing is approximately 0.34 nm. Due to the weak attraction forces between the layers, graphite develops a soft and slippery property as the weak-bonded planes can slide when acted upon by an external force. These qualities make graphite an excellent lubricant.

Like any crystalline lattice, the formation of new allotropes often involves nucleation (the initial formation of tiny seed clusters of the new structure) followed by growth. Temperature and pressure affect the nucleation and growth rates of these new structures. By manipulating temperature and pressure, we can control which allotropic form of carbon is thermodynamically favored and kinetically accessible, allowing us to create or transform different carbon allotropes. Because of this, something as simple as the ligneous bark of a tree or a dried stalk of grass can be transformed into a graphite battery or the woven carbon fiber of a Formula One race car.

A carbon life form of photosynthetic origin, such as a grass stalk, can become a diamond under the right conditions. Or, it might become an optical window for a high-power laser, a pencil lead, a fuel cell electrode, a moderator in a nuclear reactor, a solar cell, a structural nanotube reinforcement in composites, a desalinator filter, printer toner, a medicinal detoxifier, a field dressing, a rubber compound, or a cutting tool for the most rigid steel alloy. Of course, each application has attendant production costs—in finance, labor, energy, and materials—which must inform the exergy calculation. This is why biochar has emerged as a lower-exergy contender for jobs that used to be the exclusive province of diamonds, graphene, and graphite—more costly allotropes.

Today, fossil fuels are used in thousands of products: plastics to petrol, food preservation to fertilizers, and petroleum jelly to pillows. However, it took many decades, and in some cases more than a century, for most of these products to be developed and deployed. Biochar is poised to expand its product portfolio similarly, though in much less time.

One consideration in developing a particular application is whether the product being introduced to the market will have a life after death. That is, it may have additional uses beyond its intended application after its first use. This gives the product added value that can make the difference between economic viability and impracticality.

1.2 Examples of Cascades

Here are a dozen potential sequences, ranging from simple two-step cascades to more complex multi-use chains:

1. Clothing insulation → prosthetic device → tennis rackets → tires → road surfacing → erosion control.
2. Water filter → litter absorbent → compost additive → agricultural soil enhancer
3. Acoustic insulation → concrete additive → landfill cover material
4. Electronics component → activated carbon → solar refrigeration → wastewater treatment → coke reducing agent → caskets
5. Food packaging → odor absorber → pet litter additive → landscape mulch
6. Textile dye → air purifier → enzyme recovery in biofuel transesterification
7. 3D printing material → aquarium filter → hydroponics medium → biogas production additive
8. Paint additive → electromagnetic shielding → phthalate binder → road surfacing
9. Shoe insoles → mushroom growing substrate → automotive body panels
10. Biomedical implant material → industrial catalyst → energy storage
11. Supercapacitor electrode → gas storage medium → soil erosion control
12. Cosmetic ingredient → toothpaste additive → greywater treatment

These cascading use sequences are just a few examples that showcase biochar's versatility across multiple industries and applications, maximizing its value before assuming its final role as long-duration carbon sequestration.

2. Properties of Relevance

Although much peer-reviewed literature about biochar has been published over the past decade, most have focused on biochar in agriculture. More research on non-agricultural end uses is critical to identifying which particular physical, chemical, biological, and electromagnetic properties are most relevant for different uses and the optimal ranges for those properties.

Levels for heavy metals and other contaminants must be low for biochar used in soils and even lower for biochar used in animal feed. However, when biochar is used in composites, particularly when the likelihood of biochar separating from the other materials is low, higher contamination can be tolerated, as it will be sequestered within the composite. The limiting factor for impurities may have been more about their effect on the product. In graphite production, for instance, agricultural bio-waste volatiles and impurities can interfere with the synthesis process and reduce the quality of the graphitic carbon produced (Bhoi et al. 2024). Conversely, biochar that has accumulated heavy metals during water purification processes can be repurposed into electrodes for supercapacitors, stable catalyst materials, and co-catalytic systems to enhance oxidation.

Once potential benefits beyond sequestration have been prioritized, identifying the properties required to achieve those benefits should be easier. In many cases, more research may be needed. Given the emergence of non-soil uses for biochar and the greater use and scrutiny of carbon credits, new parameters are beginning to be factored in, such as the average durability of the specific biochar material and the most common end-of-life fate of biochar-enhanced materials. In certain materials such as concrete and asphalt, biochar's durability is longer than when used in soils, and the labile fraction is not subject to degradation as it is in soils (Chiaramonti et al. 2024).

These nuances are beginning to be factored into carbon credit calculations for products with shorter lifespans that are thermally treated or incinerated, substantially reducing the credit value. For example, asphalt is an emerging market for vast amounts of biochar. Since asphalt typically lasts 15 to 25 years and, in some cases, may be subject to thermal treatment at the end of its useful life, this may reduce the eligibility for carbon credits.

Shorter-duration materials remain attractive biochar options for displacing fossil fuels or mined materials. However, the expectation that the carbon credit revenues for these use cases will be the same as those for more durable materials moving forward is likely unrealistic, given the current trend by several large buyers to push for 1,000-year permanence.

3. Examples of Non-Agricultural Biochar Products

Many of these products are still in the research or early development stages. Some may still need to demonstrate commercial viability or face technical challenges that require further research and development.

3.1 Building Material

Biochar-infused Concrete

The pore structure of biochar confers added benefits on cementitious products (Murali and Wong 2024). They are lighter, improve the immobilization of noxious contaminants in sediments, have improved internal curing properties, and may self-heal micro-cracks in some circumstances by encouraging the formation of calcium carbonate (Jia et al. 2023). Biochar contributes to durability, acoustic and noise absorption, thermal insulation, and curing (Chen et al. 2023). As a filler, it increases the acoustic absorption coefficient of concrete in the 200~2000 Hz frequency range and reduces the thermal conductivity. The optimum Portland cement replacement with biochar is 1–2 wt% (by weight) for enhancing the compressive and flexural strength. Significant proportions (5 to 10 wt%) can be sustained without strength reductions. However, durability improvements at higher concentrations may result from enhanced hydration and physical filling, leading to a denser microstructure that prevents ion and water penetration. Additionally, biochar can improve resistance to sulfate attack, chloride-induced corrosion, shrinkage, and permeability in composites (Lin et al. 2023).

Fillers

A high biochar dosage (up to 60 wt%), where compressive strength is unimportant, can make carbon-negative concrete feasible. The lightweight, brittle nature of biochar reduces the carbon footprint of transportation and energy use associated with conventional fillers (Habert et al. 2020). Recently, Carbo Culture in Kerava, Finland, developed a Biographite® carbon filler that can be heated to dissipate electricity. These functionalities open the door to higher-value applications.[1]

Aggregates

Mining, crushing, and transporting some 17.5 Gt of aggregates, such as sand and gravel, produce considerable CO_2 emissions each year (Miller et al. 2018). Over-extraction of natural aggregates creates significant environmental, economic, and health issues in many countries (Ioannidou et al. 2017). By incorporating biochar as aggregates, concrete can sequester 59–65 kg CO_2 per ton and reduce production costs by up to USD$41.1 per cubic meter. Biochar-

[1] https://carboculture.com/resources/carbo-culture-wins-sprind-with-conductive-carbon-negative-concrete

enhanced concrete with 30 wt% biochar has been effectively converted into carbon-negative products (Chen et al. 2022). Marveled Designs is an artisan concrete company based in the Hudson Valley of New York. In 2024, they successfully produced a nearly carbon-neutral, 3D-printed, precast concrete with rice husk biochar as an aggregate.[2] TCHAR, a biochar company in Taiwan, is fabricating artificial coral reefs using biochar concrete.

Cement

While coarse-grained biochar can partially replace aggregate in concrete, fine-grained biochar can serve as a supplementary cementitious material. Biochar is an internal curing agent during cement hydration, aiding strength development (Askar et al. 2024). As low as 1 percent and up to 5 percent replacement of the fine aggregate in cementitious composites has been found to improve the compressive strength by approximately 10 percent (Akinyemi and Adesina 2020, Senadheera 2023). Moreover, biochar additives provide crack-arresting mechanisms at the nanoscale and microscale, thereby restricting crack growth under service loads. Biochar is a high-performance nano- and micro-reinforcement in the cementitious matrix (Khitab et al. 2021).

Mortars and Plasters

Gupta et al. reported a 7 percent reduction in the fresh density of mortar incorporating rice waste biochar at 5 percent replacement of the Portland cement (Gupta et al., 2018). A similar reduction in dry density was also reported when rice husk biochar was used as filler in mortar. 1–2 wt% biochar addition is optimal for improving the compressive strength of cement mortar. Similar effects have been observed with clay paints and plasters–acoustic and noise absorption, thermal insulation, and pollutant removal, such as smoke and mold spores. Additionally, biochar in paints and plasters may help block electromagnetic smog.

Road Surfacing

Incorporating biochar into road construction offers several advantages, including thermal insulation, mitigating the urban heat island effect, reducing noise, pollutant fixation, and capturing toxic stormwater runoff. In road construction, biochar provides strength benefits over conventional methods both in the subsurface foundation and at the surface (base and wearing course). There are early indications that, in areas that apply road salt every winter, salt combines with carbon to harden the road surface and reduce pothole formation, in addition to the already known value of biochar in reducing deicer concentration in melted snow and ice. More than 10 million tons of deicing salt is used in the USA annually, with attendant environmental issues (Kang et al. 2022).

---

[2] https://marveleddesigns.com/about-us/

Biochar will be a critical carbon removal strategy, not only because it offers co-benefits that can accelerate adoption, such as in regenerative agricultural systems, but also because it can produce many other valuable products that are part of a circular carbon economy advancing economic and social development in the coming century. Transitioning from an extractive, carbon-emitting global economy to a 21st-century circular carbon economy will require a well-thought-out, whole-systems design approach to the human-built environment.

Verde Resources has partnered with the U.S. National Center for Asphalt Technology to develop a biochar-based asphalt substitute that eliminates the need for lime stabilization in clay subgrades. The use of enzyme-inoculated biochar slashes the carbon footprint, enhances soil stabilization and moisture resistance, delivers at least 30 percent cost savings in construction, and reduces the expenses associated with sourcing and transporting traditional road materials (USSEC 2024). Verde is the first company in North America to receive carbon removal credits for biochar used in asphalt.

Biochar Catalysts

Biochar can be used as a reducing agent in direct reduction or blast furnace iron-making to convert iron oxides into metallic iron. Biochar's morphological properties (porosity and reactivity) can influence flux and slag quality in the steelmaking process (DiGiovanni et al. 2023). Biochar utilization can reduce refractory degradation and improve desulfurization and slag production (Ibitoye et al. 2024).

Carbon Fiber-reinforced Polymer (CFRP) Wraps

Carbon fiber nets can be wrapped around existing columns to prevent brittle failure during an earthquake (Bousselham and Chaallal 2004; Teng and Chen 2003; Wang and Lee 2018; Seible and Priestley 1997). This is especially important in seismic regions where reinforced concrete structures may be prone to shear failure. Biochar can produce carbon fiber with properties comparable to conventional methods (Baker and Rials 2013; Dumanli and Windle 2010; Fang et al. 2016; Meng et al. 2019).

Insulation

Biochar has low thermal conductivity (0.08 to 0.2 W/(m·K)) (Osman et al. 2024). Given the $100 billion annual market for expanded polystyrene that must be replaced with a non-fossil alternative, two start-ups have already prototyped biochar insulation. Carbon Cell has developed a spray foam material, and Elder6 has prototyped a biochar foam board. Loose rice husk biochar has been used in Korean roofing systems as an insulator, a humidity controller, and a pest deterrent.

Biochar Wallboard

Wallboard, also known as drywall or plasterboard, is ubiquitous in homes around the world. Biochar reduces thermal conductivity, decreases density while maintaining adequate strength, can meet or exceed American Society for Testing and Materials International (ASTM) standards for flexural strength at specific inclusion rates, and improves the product's recyclability. Biochar wallboard products may also offer improved moisture regulation, enhanced indoor air quality, and the potential for electromagnetic shielding.

Biochar can sequester 59–65 kg of CO_2 per ton of concrete when used as an aggregate, and a similar or higher sequestration rate may be achievable in wallboard. Assuming biochar replaces 10–20 percent of gypsum in wallboard, and considering the market size, the potential for carbon sequestration could be in the range of millions of tons of CO_2e per year. There is end-of-life recyclability as a soil amendment. The challenges may include ensuring cost competitiveness with traditional gypsum boards and addressing regulatory or building code barriers.

To estimate the potential scale, if biochar replaced 10 percent of gypsum in 20 percent of the global wallboard market and sequestered 60 kg CO_2e per ton, the annual sequestration potential could be in the range of 1–2 million tons of CO2e per year, scaling up as market penetration increases. This is a rough estimate and would require more detailed market analysis and life cycle assessments to be refined. However, it illustrates that biochar in wallboard has significant potential to contribute meaningfully to carbon sequestration while providing additional performance benefits.

Biochar SIPs

Adding biochar to Structural Insulated Panels (SIPs) enhances performance and fire-resistant properties, and reduces overall panel weight without compromising strength, making transportation and installation easier (Li et al. 2023). SIPs are known for their superior insulation properties, but biochar could further enhance them. This could lead to even greater energy efficiency in buildings, reducing heating and cooling costs. Biochar can absorb up to five times its weight in water. In SIPs, this quality could regulate humidity, improve indoor air quality, and reduce the risk of mold growth. The porous structure of biochar can improve the acoustic insulation of SIPs, enhancing their noise-reduction performance. While there are challenges to overcome, incorporating biochar into SIPs presents an opportunity for these already eco-friendly building materials (SIPA 2023).

Biochar Roofing Materials

Incorporating 30–40 percent by volume biochar into green roof substrates can significantly improve water and nutrient retention and reduce the overall weight of the roofing system. In asphalt roofing, biochar-asphalt composites have demonstrated improved mechanical properties, insulation, and reduced weight. A project at the Rochester Institute of Technology explored the creation of roof tiles using a mixture of approximately 30 percent by volume biochar (10 percent by weight), combined with cement, sand, and water.[3]

Biochar-based Flooring

We have seen that replacing 1 wt% of the cement with biochar can increase concrete compressive strength by up to 20% after 28 days of curing (Zhang et al. 2022). Additionally, the biochar-modified concrete exhibited enhanced resistance to chloride ion penetration, suggesting improved durability in aggressive environments (Gupta et al. 2022). In composite flooring materials, incorporating 24 wt% biochar into high-density polyethylene (HDPE) composites resulted in a 35% reduction in water absorption compared to conventional wood-filled composites (Väisänen et al. 2026). This improvement in moisture resistance could lead to more durable and stable flooring materials, particularly in humid environments.

Another promising avenue for biochar flooring materials is in the development of bio-based resins and adhesives. The addition of 2 wt% biochar to the PU coating improved its scratch resistance by 25% and increased its hardness by 30% (Uram et al. 2021). In the context of sustainable building insulation, when applied to flooring systems, incorporating 30 vol% of biochar into a gypsum matrix reduced thermal conductivity by 51% compared to pure gypsum.

3.2 Filtration

Stormwater Runoff

Biochar is an effective and efficient water filtration medium. The media may later be cascaded for its agronomic benefits depending on contamination (Werner et al. 2018). Activated biochars offer an effective solution for the removal of Per- and Polyfluoralkyl substances (PFAS, also called "forever chemicals") (Yadav et al. 2022). Treated biochar filters, such as with TiO_2, can destroy PFAS and regenerate the biochar for reuse (Xu et al. 2020). Simultaneous removal and repurposing of other contaminants, such as phosphorus, nutrients, heavy metals, and dissolved organic matter, offer many practical applications, including slow-release phosphorus fertilizers, additives for concrete, and novel adsorbents for heavy-metal removal (Wang et al. 2025).

Biochar Desalination Membranes

Recent research has demonstrated the potential of biochar-based solar interfacial evaporators for efficient seawater desalination. These membranes demonstrate high mechanical strength, a porous structure with uniform biochar particle distribution, increased water flux compared to polymer-only membranes, improved hydrophilicity and surface free energy, promising adsorption

[3] https://wiki.opensourceecology.org/wiki/Biochar_Roof_Tiles

capacities for contaminants such as dyes, and high retention of bacteria (e.g., *E. coli*). Biochar acts as a photothermal material, absorbing solar energy and converting it to heat. Its porous structures allow for efficient water transport and vapor generation.

Biochar-based evaporators show high salt rejection rates and can meet drinking water quality standards. When tested under 1 sun illumination, a biochar evaporator achieved a remarkable water evaporation rate of 3.173 kg m^{-2} h^{-1}, with a solar-to-vapor conversion efficiency of 100 percent. The biochar evaporator displayed good resistance to salt crystallization and produced potable drinking water from seawater, alkaline wastewater, and organic-polluted water (Zhang et al. 2021).

Notably, the evaporator demonstrated excellent desalination performance, removing over 99.9% of major ions (Na^+, K^+, Mg^{2+}, Ca^{2+}) from seawater. The resulting distillate met World Health Organization drinking water standards. Furthermore, the evaporator showed stable performance over multiple cycles, maintaining high efficiency without observable salt accumulation or structural degradation.

HAB Filtration

Biochars can destroy harmful algal blooms (HABs) by increasing reactive oxygen species (Moore et al. 2023). Iron-modified biochar is up to 98 percent effective at removing phosphorus from wastewater across a range of pH and other conditions, compared to the more costly, non-renewable MgO typically used in wastewater treatment. Biochar can also effectively mitigate Ca, K, Mg, and Na in dairy sludge effluent (Vogler 2023).

Biochar Industrial Filters

Biochar effectively removes metal vapors, particularly elemental mercury (HgO), as well as acidic gases such as H_2S, SO_2, and CO_2, ozone, nitrogen oxides (NOx), and volatile organic compounds (VOCs) (Gwenzi et al. 2021). In water treatment applications, biochar-based filters, especially in mixed filter beds, achieved higher removal efficiencies for turbidity, copper, and coliform bacteria than traditional sand filters (García-Ávila et al. 2023).

Tripathi et al. (2024) reported on the growing interest in biochar for wastewater treatment. Their research found that Miscanthus biochar filters achieved significantly higher chemical oxygen demand (COD) removal (74±18%) than sand filters (61±12%). Additionally, the biochar filters demonstrated superior *E. coli* removal, with a reduction of 1.35±0.27 log units compared to 1.18±0.31 log units for sand filters. Nanosheet-like Fe_2O_3 biochar has demonstrated the ability to remove 97% of tetracycline within 80 minutes. The degradation mechanism emphasizes the significant role of hydroxyl radicals (OH) in degrading pharmaceutical and antibiotic pollutants (Wu et al. 2024). A biochar-loaded Fe-Cu layered double hydroxide is an innovative method for degrading Penicillin G (PG), a widely used antibiotic (Liu et al. 2024).

3.3 Energy

Electronics

Biochar has demonstrated performance comparable to that of graphite in specific applications, such as an electrode material in microbial electrochemical technologies (METs). Biochar electrodes can achieve current densities comparable to those of graphite while being more cost-effective and environmentally friendly (Gorrazzi et al. 2023). Temperature, pressure, feedstock, catalyst, and other parameters must be carefully controlled to obtain the desired product. For instance, high-temperature biochar (formed at 800 to 1000 °C) has been shown to perform comparably to graphite in METs, while low-temperature biochar (formed at 600 °C) generates about half the current (Banek et al. 2022).

Carbon Fuel Cells

Combining a fuel cell system and biomass was suggested by Grove in 1839 (Appleby 1990). A carbonaceous solid fuel is heated to generate electricity directly from its chemical energy, without gasification. The direct carbon fuel cell (DCFC) uses carbon as the anode and operates at 700–900 °C. DCFC has a greater attainable efficiency (80 percent) for power generation than other fuel cells, such as hydrogen (Jafri et al. 2018). The feedstock for fueling DCFC is a solid carbon-rich fuel, such as coal, biomass, or organic waste. The maximum power density of biochar is 60–70 percent of that of coal, making it a viable alternative to fossil fuels (Ahn et al. 2013). The exhaust of this system is carbon dioxide, as indicated in the overall reaction:

Anode reaction: $C + 2CO_3 \rightarrow 300r + 4e$
Cathode reaction: $O_2 + 2CO_2 + 4e \rightarrow 2CO_2$
Overall reaction: $C + O_2 \rightarrow CO_2$

Microbial Fuel Cells

Biochar has been used in the preparation of electrodes, binding materials, and proton exchange membranes for microbial fuel cells (MFCs), where fabrication, cathode, and membrane production account for a significant fraction of the total cost. MFCs are electrochemical devices that use a biological system to generate electrons via the anaerobic oxidation of a substrate (Lee et al. 2021). Heavy metals and graphite are the common materials used, but they are costly and generate secondary toxic waste. Biochar can be used in electrodes, in the fabrication of the separating membrane, and as a low-cost, efficient filling material in the place of graphite granules and granule-activated carbon. Biochar produced at higher temperatures is suitable for cathodic electrodes due to its high oxygen-reducing nature and surface area. For anode electrodes, biochar-doped graphite (graphite + 0.5 mg cm^{-2} biochar) shows an increase in power density and a decrease in internal resistance (Ahuja et al. 2024).

Green Hydrogen

Biohydrogen is a type of green hydrogen derived from pyrolyzed biomass. It offers potential advantages over traditional hydrogen production methods. It can be carbon-neutral or even carbon-negative when produced from feedstocks like waste and manure. When made with high-temperature, industrial-scale, commercial-grade biochar systems, it has 25 percent of the capital cost of electrolysis, uses two-thirds less water, and requires no external energy source. According to the European Biogas Association, such biochar systems can generate twice as much hydrogen per unit of energy input. It is compact, modular, scalable, and can be made from various biomass waste, allowing for co-location close to the point of use and reducing transport costs. Biochar nanocarbon low-pressure hydrogen storage reduces the need for high-cost compression in production, transport, and storage (Bertasini et al. 2023).

Batteries and Supercapacitors

The hierarchical porous structure of biochar with heteroatom surface functionalities is a desirable feature for supercapacitor application (Teo et al. 2016). Electrodes have even been made from pyrolyzed shrimp shells (Huang et al. 2021). For metal ion and Lithium battery applications, biochar derived from wood chips, oil palm stone, coconut shell, rice straw, lotus seedpod, wheat bran, cotton, banana peel, popcorn, bamboo, hair, loofah sponge, and lignin has been successfully substituted for metallic components, providing low cost, abundance, and renewability (Gao et al. 2018). A high surface area, a micropore-dominant structure, and O-rich functional groups are the primary assets for hydrogen storage in biochar. It can be customized to enhance its structural and electrochemical properties by imparting porosity, increasing surface area, enhancing graphitization, or modifying surface functionalities through heteroatom doping. All these biochar features are highly desired in energy storage (Rawat et al. 2023).

Biochar Solar Panels

Most photoinduced semiconductors, including TiO_2, are considered undesirable due to their large band gaps and limited response to visible light. Biochar's porous structure and vast surface area make it an excellent platform for the deposition of photocatalytic nanoparticles, promoting higher CO_2 conversion rates. As the photocatalyst absorbs light and generates electrons, biochar is an efficient mediator, absorbing and delivering these electrons to the photocatalyst's surface (Mohtaram et al. 2024).

EMI Shielding Materials

As homes become increasingly reliant on electronic devices and smart technologies, electromagnetic interference (EMI), or "EM smog," has become a growing concern for homeowners. Continuous exposure to electromagnetic radiation, whether from Wi-Fi routers, power lines, or household appliances, can lead to discomfort or health concerns for some individuals.

A study published in *Composites Science and Technology* (Jiang et al. 2019) combined biochar with epoxy and polystyrene matrices, showing an electromagnetic frequency (EMF) shielding effectiveness of 25-30 dB in the frequency range of 8 to 12 GHz (X-band). Another study investigated the use of biochar from coconut shells and demonstrated shielding against low-frequency EM waves (Raghavan et al. 2020). When mixed with conductive polymers, the composite achieved a shielding effectiveness of up to 40 dB, indicating the potential of biochar for shielding against both high- and low-frequency radiation. Biochar modified at temperatures above 900 °C exhibited significant improvements in electrical conductivity and dielectric properties. These high-temperature modifications enable biochar to better absorb and reflect electromagnetic waves, making it a viable material for EM shielding (Wang et al. 2021). Biochar is a competitive replacement for conventional metal-based shields due to its light weight, thermal stability, cost-effectiveness, and renewable origin.

A cost-effective strategy is a layered approach, using biochar-based paints or plasters as a base and applying metallic or conductive additives in specific areas. This minimizes the need for expensive materials while providing effective EM shielding in key areas, such as bedrooms, living rooms, or home offices. The integration of biochar for cost, effectiveness, and consumer appeal provides a solid foundation, while hybrid products that incorporate metallic and conductive additives offer enhanced performance.

Carbon Nanotubes

Waste plastics can produce hydrocarbon nanotubes (CNTs) when pyrolyzed. With special properties derived from curling graphene sheets, CNTs are widely used in electronic information and energy storage, analytical chemistry, and aerospace. Waste plastics can be used as carbon sources to prepare CNTs by chemical vapor precipitation, yielding hydrogen as a byproduct for fuel or in fuel cells. While it is difficult to characterize pyrolyzed waste plastics as biochar, it is a type of capture that can fit into a carbon cascade.

Biochar-based Flexible Electronics

Biochar-polymer composite fibers with electrical conductivity and flexibility have shown potential for smart textile applications and as alternatives to conventional electric wires (Łukaszewicz et al. 2021). Notably, biochar composites have exhibited electrical properties comparable to those of carbon nanotubes, with 20 wt% biochar composites demonstrating performance similar to that of 4 wt% carbon nanotube composites.

The realm of 3D printing has also seen advancements in biochar-based materials. Researchers have developed biochar composites with thermoplastic polymers commonly used in additive manufacturing, opening avenues for three-dimensional electronics and printed electronic applications. Companies such as Carboganic, located in the BlueChem incubator in Antwerp, hope to revolutionize 3D-printable compounds for FDM (Fused Deposition Modeling) and filaments in the production of furniture, boats, bikes, houses, prosthetics, cables, drones, and more.[4]

Biochar derived from biomass waste has enabled the fabrication of flexible, biodegradable electroluminescent devices using biochar/polymer composite films that exhibit good mechanical flexibility, electrical conductivity, and light-emitting properties (Wang et al. 2021). Screen-printed electrodes (SPEs) have been successfully produced in printed electronics (Cancelliere et al. 2022), and flexographic inks using pine char as a substitute for carbon black have been used to print electronic components (Goh et al. 2021). A stabilized biochar-based ink derived from fruit peels, applied to various flexible substrates, may have applications in wearable electronics and flexible displays (Yuan et al. 2021).

3.4 Consumer Products

Biochar-infused Fabrics

Biochar fibers are increasingly incorporated into textiles, including sportswear, to enhance performance and sustainability. Bamboo-based biochar is one of the oldest and most common of these. It has a more matte look and linen feel than silkier bamboo fabrics. When used in apparel, bamboo biochar has established itself for several benefits, including trapping odor-causing molecules to help keep garments fresher for longer; absorbing moisture to help keep the wearer dry during physical activity; temperature-regulating properties; softness; and UV protection.[5]

While bamboo is currently the most widely used feedstock, other feedstocks are also being explored for textile applications, such as biochar derived from textile waste, which can be added to cotton fabrics to improve performance and add functional properties (Çay et al. 2020). TCHAR has recently debuted an artificial leather with biochar in Taiwan for use in furniture and car interiors. Ethical brands have incorporated eco-friendly fabrics like recycled polyester, bamboo viscose, certified organic cotton, and Tencel, as well as recycled materials like PET and nylon, to make sustainable athletic wear.

Carbon Caskets

While the combination of biochar and coffins/burials dates at least to Justis von Liebig in 1873 (Brock 2002), the practice has not been widely adopted. However,

---

[4] https://www.carboganic.eu/applications

[5] https://www.sportcasuals.com/news/trending-activewear-fabrics; https://www.eco-stylist.com/10-sustainable-activewear-brands/

it is conceivable that a product such as biochar paper-mâché, woven fiber, or polymer composite casket could be developed, given the ubiquity of biochar producers and the emergence of eco-friendly burial product manufacturers, such as Natural Burial Company, Mourning Dove Studio (cardboard coffins), and Passages International (woven fiber caskets).

Sporting Goods

While not specifically using biochar, Wilson Sporting Goods has developed an eco-conscious tennis racket series called "The Limited-Edition Naked Series". These rackets use sustainable materials, such as castor bean oil-derived plastics and biodegradable grips, reflecting a trend toward more environmentally friendly materials in sporting goods. Biochar has already been used in golf courses, football pitches, and other sports fields to improve soil health, reduce irrigation needs, and enhance turf performance. We can expect to see it used in sporting goods made from polymer composites, potentially improving mechanical, thermal, and electrical properties.

At the 2025 Snowboard, Freestyle and Freeski World Championships, riders on biochar composite boards took the first and second positions in the mixed team snowboard parallel slalom event. The overall 2024–25 Alpine Snowboarding World Cup went to a biochar polymer board rider. One of the advantages was the reduction of electrostatic charging caused by the friction between the snowboard and the more or less dry snow, due to the biochar's electrical conductivity and electron buffering. Adding biochar to the polymer glass deck improved the board's ability to absorb energy during hard carving turns and then release energy on rebounding at the end of the turn, boosting acceleration. The biochar improved the board's ability to store energy when it bends and shed it when released. Vibration is transformed from speed loss (energy loss to heat and noise) to speed gain as the nano-sized biochar particles act like a capacitor within the composite material, capturing and returning energy in response to the board's bending stage (Schmidt 2025).

Biochar Tires

Biochar shows potential as a partial replacement for carbon black in vulcanized rubber compounds used in tires. Incorporating biochar into rubber composites improves tensile strength, thermal stability, and electrical conductivity (Peterson 2020). Compared to petroleum-based carbon black, biochar can be produced from waste biomass without compromising critical attributes (Lepak-Kuc et al. 2021). Key tests, including rheometry and differential scanning calorimetry, confirmed that biochar can positively influence cross-linking in rubber, but its susceptibility to aging requires further improvement (Miedzianowska-Masłowska et al. 2024).

Solar Refrigeration

These same qualities used for solar evaporators make biochar an ideal medium for evaporative cooling. In 2009, a team of engineering students from Michigan State University working in Quetzaltenango, Guatemala, built a vaccine refrigerator capable of maintaining a temperature between 2 °C and 8 °C using passive solar energy to evaporate ethanol in a biochar matrix, condense it, and cycle the coolant back through. The biochar used did not need to be agricultural-grade; it could come from a variety of feedstocks.

Using charcoal, wire mesh, and a water tank, the women of the Bidii Farmers Group in the arid Kambi Sheikh Village, Isiolo County, Kenya, built a walk-in cooler to store their produce for market. The cooler had a capacity of 20 crates of produce. A network of perforated water pipes directed water from an overhead tank through biochar-insulated roof and walls. As the water continuously dripped through the biochar, evaporated, and condensed for collection, the temperature inside dropped to 8 °C (46 °F).

Kitty Litter

USDA researchers have improved an earlier biobased cat litter formulation by adding biochar from soybean hulls and a starch-based ingredient. Compared with four commercial biobased cat litters, this formulation was superior for odor and dust reduction, while having hydration capacity and clumping ability equal to or better than those of the commercial litters (Vaughn et al. 2020, 2023).

3.5 Future Prospects

As industrial designers and researchers continue to explore this field, we may witness a shift towards more climate-friendly and economically viable biochar processes and components, potentially reshaping many industrial processes. Future applications include: sporting goods such as snowboards, tennis rackets and sportswear; aerospace composites (Aboughaly et al. 2023); automotive body panels (Das et al. 2021); tires (Peterson 2020, Lepak-Kuc et al. 2021; Miedzianowska-Masłowska et al. 2024); ship hull coatings (Marchesini 2023); solar refrigeration; wound dressings and drug delivery systems (Zhuo et al. 2023; Bernardes et al. 2021; Kim et al. 2023); Diethyl phthalate (DEP) binder (Agilandeswari et al. 2024); prosthetic devices (Akter et al. 2022; Garg et al. 2024); kitchenwares (Ak et al. 1994); enzyme recovery in biofuel transesterification catalysts (Tan et al. 2024); coke reducing agents (Echterhof 2021); slag foaming agents (Chiaramonti et al. 2024); and even caskets.

4. Conclusions and Recommendations

According to a 2024 G20 Task Force policy brief (Khanna and Pangelinan 2024): Especially in developing countries with dominant rural populations and heavy dependence upon natural resources, (social enterprise) clusters holistically transit the local micro-economies towards decarbonization while delivering triple-bottom-line benefits.

In *Burn: Igniting a New Carbon Drawdown Economy to End the Climate Crisis* (Bates and Draper 2020), the authors described four rural development models underway, which we dubbed "Cool Labs". Each started from the bottom and built a regenerative, circular economy based upon local assets—human and natural. There are many more examples today. These microeconomic laboratories use biochar to produce transformative products. They are constructing a circular carbon economy at the village, district, and bioregion scales (Khanna et al. 2024). With the adoption of the Paris Agreement in December 2015, 196 UN member nations committed themselves to holding the global average temperature to 1.5 to 2 degrees above pre-industrial levels. The world is now well on track to exceed the Paris temperature goals in the early 2030s, if not sooner. Scaling carbon removal is vital to humanity's future. Low-cost, locally-based, economically productive, and socially responsible small enterprises are a winning strategy. The range of new products and services biochar offers may be just the fuel they need to ignite their aspirations.

Biochar is a key climate strategy, not only because it delivers co-benefits that can accelerate adoption, such as in regenerative agricultural systems, but also because it can produce many non-agricultural products and services that are part of a circular carbon economy. In an era of cascading crises caused by accelerating climate change, biochar provides cascading opportunities for enterprise and innovation. Most studies of biochar in the past decade have focused on agricultural applications. More research on non-agricultural end uses is critical to scaling this climate solution in the future and fully utilizing its many economic and social benefits.

References

Aboughaly, M., Babaei-Ghazvini, A., Dhar, P., Patel, R., and Acharya, B. 2023. Enhancing the potential of polymer composites using biochar as a filler: A review. *Polymers* 15(19): 3981.

Agilandeswari, P., Venkateshbabu, S., Rajamohan, N., Sarojini, G., and Rajasimman, M. 2024. Synthesis and functional mechanism evaluation of novel polymer-graphene oxide-biochar nanocomposite for efficient removal of diethyl phthalate. *Environmental Research* 120032.

Ahn, S.Y., Eom, S.Y., Rhie, Y.H., Sung, Y.M., Moon, C.E., Choi, G.M. et al. 2013. Utilization of wood biomass char in a direct carbon fuel cell (DCFC) system. *Applied Energy* 105: 207–216.

Ahuja, V., Palai, A.K., Kumar, A., Patel, A.K., Farooque, A.A., Yang, Y. et al. 2024. Biochar: Empowering the future of energy production and storage. *Journal of Analytical and Applied Pyrolysis* 177: 106370. https://doi.org/10.1016/j.jaap.2024.106370

Ak, N.O., Cliver, D.O., and Kaspar, C.W. 1994. Cutting boards of plastic and wood contaminated experimentally with bacteria. *Journal of Food Protection* 57(1): 16–22.

Akinyemi, B.A., and Adesina, A. 2020. Recent advancements in the use of biochar for cementitious applications: A review. *Journal of Building Engineering* 32: 101705.

Akter, M., Uddin, M.H., and Tania, I.S. 2022. Biocomposites based on natural fibers and polymers: A review on properties and potential applications. *Journal of Reinforced Plastics and Composites* 41(17–18): 705–742.

Appleby, A.J. 1990. From Sir William Grove to today: Fuel cells and the future. *Journal of Power Sources* 29(1–2): 3–11.

Askar, A. et al. 2024. Biochar integration in precast concrete a critical appraisal of sustainability and performance enhancement. Civil Engineering Research in Ireland 2024 (CERI2024) Conference.

Baker, D.A., and Rials, T.G. 2013. Recent advances in low-cost carbon fiber manufacture from lignin. *Journal of Applied Polymer Science* 130(2): 713–728.

Banek, N.A., McKenzie, K.R., Abele, D.T., and Wagner, M.J. 2022. Sustainable conversion of biomass to rationally designed lithium-ion battery graphite. *Scientific Reports* 12(1): 8080.

Bates, A., and Draper, K. 2020. *Burn: Igniting a New Carbon Drawdown Economy to End the Climate Crisis.* Chelsea Green Publishing.

Bernardes, B.G., Del Gaudio, P., Alves, P., Costa, R., García-González, C.A., and Oliveira, A.L. 2021. Bioaerogels: Promising nanostructured materials in fluid management, healing and regeneration of wounds. *Molecules* 26(13): 3834.

Bertasini, D., Battista, F., Rizzioli, F., Frison, N., and Bolzonella, D. 2023. Decarbonization of the European natural gas grid using hydrogen and methane biologically produced from organic waste: A critical overview. *Renewable Energy* 206: 386–396.

Bhoi, R., Patro, A., and Sarkar, A. 2024. Sustainable transformation of agricultural waste into value-added end products through thermochemical approach and end product characteristics. *In: Agricultural Waste to Value-Added Products: Bioproducts and its Applications*, pp. 1–25. Singapore: Springer Nature Singapore.

Bousselham, A., and Chaallal, O. 2004. Shear strengthening reinforced concrete beams with fiber-reinforced polymer: Assessment of influencing parameters and required research. *ACI Structural Journal* 101(2): 219–227.

Brock, W.H. 2002. *Justus von Liebig: The Chemical Gatekeeper.* Cambridge University Press.

Cancelliere, R., Cianciaruso, M., Carbone, K., and Micheli, L. 2022 Biochar: A sustainable alternative in the development of electrochemical printed platforms. *Chemosensors* 10(9): 344.

Çay, A., Yanık, J., Akduman, Ç., Duman, G., and Ertaş, H. 2020. Application of textile waste derived biochars onto cotton fabric for improved performance and functional properties. *Journal of Cleaner Production* 251: 119664.

Chen, L., Zhou, T., Yang, J., Qi, J., Zhang, L., Liu, T. et al. 2023. A review on the roles of biochar incorporated into cementitious materials: Mechanisms, application and perspectives. *Construction and Building Materials* 409: 134204.

Chen, L., Zhang, Y., Wang, L., Ruan, S., Chen, J., Li, H. et al. 2022. Biochar-augmented carbon-negative concrete. *Chemical Engineering Journal* 431: 133946.

Chiaramonti, D., Lehmann, J., Berruti, F., Giudicianni, P., Sanei, H., and Masek, O. 2024. Biochar is a long-lived form of carbon removal, making evidence-based CDR projects possible. *Biochar* 6(1): 81.

Das, C., Tamrakar, S., Kiziltas, A., and Xie, X. 2021. Incorporation of biochar to improve mechanical, thermal and electrical properties of polymer composites. *Polymers* 13(16): 2663.

DiGiovanni, C., Delin Li, Ka Wing Ng, and Xianai Huang. 2023. Ranking of injection biochar for slag foaming applications in steelmaking. *Metals* 13(6): 1003.

Dumanli, A.G., and Windle, A.H. 2010. Carbon fibers from cellulosic precursors: A review. *Journal of Materials Science* 45(16), 4258–4268.

Echterhof, T. 2021. Review on the use of alternative carbon sources in EAF steelmaking. *Metals* 11(2): 222.

Fang, H., Wang, J., Zhan, H. et al. 2016. Carbon fibers prepared from lignin/polyacrylonitrile composites. *Materials Letters* 165: 164–166.

Gao, M., Pan, S.-Y., Chen, W.-C., and Chiang, P.-C. 2018. A cross-disciplinary overview of naturally derived materials for electrochemical energy storage. *Materials Today Energy*, 7: 58–79, 10.1016/j.mtener.2017.12.005

García-Ávila, F., Galarza-Guamán, A., Barros-Bermeo, M., Alfaro-Paredes, E.A., Avilés-Añazco, A., and Iglesias-Abad, S. 2023. Integration of high-rate filtration using waste-derived biochar as a potential sustainable technology for drinking water supply. *Biochar* 5(1): 62.

Garg, S., Surendran, V.S.B., and Avanthi, A. 2024. Recent trends and applications of biochar and nanoparticles from plant biomass. *In: Medicinal and Aromatic Plants: Current Research Status, Value-Addition to Their Waste, and Agro-Industrial Potential* (Vol II), pp. 203–216. Cham: Springer Nature Switzerland, 2024.

Goh, Y., Lauro, S., Barber, S.T., Williams, S.A., and Trabold, T.A. 2021. Cleaner production of flexographic ink by substituting carbon black with biochar. *Journal of Cleaner Production* 324: 129262.

Gorrazzi, S.A., Massazza, D., Pedetta, A., Silva, L., Prados, B., Fouga, G. et al 2023. Biochar as a substitute for graphite in microbial electrochemical technologies. *RSC Sustainability* 1(5): 1200–1210.

Gupta, S., Tulliani, J.M., and Kua, H.W. 2022. Carbonaceous admixtures in cementitious building materials: Effect of particle size blending on rheology, packing, early-age properties and processing energy demand. *Science of the Total Environment* 807: 150884.

Gupta, S., Kua, H.W., and Pang, S.D. 2018. Biochar-mortar composite: Manufacturing, evaluation of physical properties and economic viability. *Construction and Building Materials* 167: 874–889.

Gwenzi, W., Chaukura, N., Wenga, T., and Mtisi, M. 2021. Biochars as media for air pollution control systems: Contaminant removal, applications and future research directions. *Science of the Total Environment* 753: 142249.

Habert, G., Miller, S.A., John, V.M., Provis, J.L., Favier, A., Horvath, A. et al. 2020. Environmental impacts and decarbonization strategies in the cement and concrete industries. *Nature Reviews Earth & Environment* 1(11): 559–573.

Huang, S., Ding, Y., Li, Y., Han, X., Xing, B., and Wang, S. 2021. Nitrogen and sulfur co-doped hierarchical porous biochar derived from the pyrolysis of mantis shrimp shell for supercapacitor electrodes. *Energy Fuels* 35: 1557–1566, 10.1021/acs.energyfuels.0c04042

Ibitoye, S.E., Loha, C., Mahamood, R.M., Jen, T.-C., Alam, M.J., and Sarkar, I. 2024. An overview of biochar production techniques and application in iron and steel industries. *Bioresources and Bioprocessing* 11(1): 65.

Ioannidou, D., Meylan, G., Sonnemann, G., and Habert, G. 2017. Is gravel becoming scarce? Evaluating the local criticality of construction aggregates. *Resources, Conservation and Recycling* 126: 25–33.

Jafri, N., Wong, W.Y., Doshi, V., Yoon, L.W., and Cheah, K.H. 2018. A review on production and characterization of biochars for application in direct carbon fuel cells. *Process Safety and Environmental Protection* 118: 152–166.

Jia, Y.I., Li, H., He, X., Li, P., and Wang, Z. 2023. Effect of biochar from municipal solid waste on mechanical and freeze–thaw properties of concrete. *Construction and Building Materials* 368: 130374.

Jiang, F., Cui, Z., Zhang, X. et al. 2019. Biochar-polymer composites for EMI shielding: Effectiveness and mechanism. *Composites Science and Technology* 182: 107707. https://doi.org/10.1016/j.compscitech.2019.107707.

Kang, Dong Hee, Al Tameemi, R., Lopez, S., and Qian, X. 2022. Evaluation of deicing salt immobilization in biochar applied BMPs. *In: World Environmental and Water Resources Congress 2022*, pp. 87–94.

Khanna, S., and Pangelinan, J.J. 2024. Promoting women enterprise clusters as agents of change in a just green transition of G20 Economies, Task Force O2: Sustainable climate action and inclusive just energy transitions. *T20 Brasil 2024 Policy Brief*, August 2024.

Khanna, S., Hari Krishna, D., Singh, S., Gupta, M., Sah, S. and Aparajita M. Matthew. 2024. Harnessing local aspirations for green and inclusive livelihoods: Establishing the impact of rural enterprise clusters in India. *The Society for Technology and Action for Rural Advancement* (*TARA*).

Khitab, A., Ahmad, S., Khan, R.A., Arshad, M.T., Anwar, W., Tariq, J. et al. 2021. Production of biochar and its potential application in cementitious composites. *Crystals* 11(5): 527.

Kim, Y., Patel, R., Kulkarni, C.V., and Patel, M. 2023. Graphene-based aerogels for biomedical application. *Gels* 9(12): 967.

Lee, S.M., Lee, H.-J., Kim, S.H., Suh, M.J., Cho, J.Y., Ham, S. et al. 2021. Engineering of Shewanella marisflavi BBL25 for biomass-based polyhydroxybutyrate production and evaluation of its performance in electricity production. *International Journal of Biological Macromolecules* 183: 1669–1675, 10.1016/j. ijbiomac.2021.05.105

Lepak-Kuc, S., Kiciński, M., Michalski, P.P., Pavlov, K., Giorcelli, M. and Bartoli, M. 2021. Innovative biochar-based composite fibres from recycled material. *Materials* 14(18): 5304.

Li, Huanyu, Wang, L., Zhang, Y., Yang, J., Tsang, D.C.W. and Viktor Mechtcherine. 2023. Biochar for sustainable construction industry. *In: Current Developments in Biotechnology and Bioengineering*, pp. 63–95. Elsevier.

Lin, X., Li, W., Guo, Y., Dong, W., Castel, A. and Wang, K. 2023. Biochar-cement concrete toward decarbonisation and sustainability for construction: Characteristic, performance and perspective. *Journal of Cleaner Production* 138219.

Liu, Z., Wang, G., Xu, T., Deng, N., Xie, H. and Zhang, X. 2024. Visible-light-driven peroxydisulfate activation by biochar-loaded Fe-Cu layered double hydroxide for penicillin G degradation: Performance, mechanism and application potential. *Environmental Research* 120043.

Łukaszewicz, S., Kowalczyk, M., Michałowski, P.P., Pakuła, K., Giorcelli, M., Bartoli, M. et al. 2021. Innovative biochar-based composite fibres from recycled materials. *Materials* 14(18): 5215.

Marchesini, J., Perondi, D., Marocco, M.V., Pellizzoni, V.G., Carberlon, G.O., Godinho, M. et al. 2023. Production and application of biochar in a UV radiation-curable epoxy paint as a substitute for graphite. *Journal of Coatings Technology and Research* 20(3): 957–971.

Masson-Delmotte, V., Zhai, P., Pörtner, H.O., Roberts, D., Skea, J., Shukla, P.R. et al. 2019. Global warming of 1.5 °C. *An IPCC Special Report on the Impacts of Global Warming of 1.5 °C, Summary for Policymakers.*

Meng, L., Dou, S., Yang, W. et al. 2019. Lignin-based carbon fibers: Carbonization and characterization. *Carbon* 147: 401–408

Miedzianowska-Masłowska, J., Masłowski, M., Delekta, M., and Strzelec, K. 2024 Waste management of straw to manufacture biochar: An alternative reinforcing filler for natural rubber biocomposites. *Industrial Crops and Products* 222: 119629.

Miller, S.A., John, V.M., Pacca, S.A., and Horvath, A. 2018. Carbon dioxide reduction potential in the global cement industry by 2050. *Cement and Concrete Research* 114: 115–124.

Mohtaram, S., Mohtaram, M.S., Sabbaghi, S., You, X., Wu, W., and Golsanami, N. 2024. Enhancement strategies in CO_2 conversion and management of biochar supported photocatalyst for effective generation of renewable and sustainable solar energy. *Energy Conversion and Management* 117987.

Moore, J., Jayakumar, A., Soldatou, S., Mašek, O.J., Lawton, L.A., and Christine Edwards. 2023. Nature-based solution to eliminate cyanotoxins in water using biologically enhanced biochar. *Environmental Science & Technology* 57(43): 16372–16385.

Murali, G., and Wong, L.S. 2024. A comprehensive review of biochar-modified concrete: Mechanical performance and microstructural insights. *Construction and Building Materials* 425: 135986.

Osman, A.I., Farghali, M., Dong, Y., Kong, J., Yousry, M., Rashwan, A.K. et al. 2024. Reducing the carbon footprint of buildings using biochar-based bricks and insulating materials: A review. *Environmental Chemistry Letters* 22(1): 71–104.

Peterson, S.C. 2020. Coppiced biochars as partial replacement of carbon black filler in polybutadiene/natural rubber composites. *Journal of Composites Science* 4(4): 147.

Raghavan, A., Balakrishnan, S., Kumar, M.V. et al. 2020. Electromagnetic interference shielding effectiveness of coconut-shell biochar reinforced polymer composites. *Materials Letters* 271: 127833.

Rawat, S., Wang, C.T., Lay, C.H., Hotha, S., and Bhaskar, T. 2023. Sustainable biochar for advanced electrochemical/energy storage applications. *Journal of Energy Storage* 63: 107115.

Schmidt, H.P., 2025. Biochar wins World Cup. *The Biochar Journal* ISSN 2297-1114.

Seible, F., Priestley, M.J.N., Hegemier, G.A., and Innamorato, D. 1997. Seismic retrofit of RC columns with continuous carbon fiber jackets. *Journal of Composites for Construction* 1(2): 52–62.

Senadheera, S.S., Gupta, S., Kua, H.W., Hou, D., Kim, S., Tsang, C.W. et al. 2023. Application of biochar in concrete – A review. *Cement and Concrete Composites* 105204.

Structural Insulated Panel Association (SIPA). Sep 12, 2023. Redefining Sustainable Construction with Structural Insulated Panels.

Tan, Z., Chen, G., Ma, X., Ge, F., Zhao, Y., and Li, A. 2024. Carbon-based magnetic nanoparticle utilizing nano-biochar as core and its immobilizing lipase for biodiesel preparation. *Industrial Crops and Products* 222: 119693.

Teng, J.G., Chen, J.F., Smith, S.T., and Lam, L. 2003. Behavior and strength of FRP-strengthened RC structures: A state-of-the-art review. *Composites Part B: Engineering* 34(6): 625–642.

Teo, E.Y.L., Muniandy, L., Ng, E.-P., Adam, F., Mohamed, A.R., Jose, R. et al. 2016. High surface area activated carbon from rice husk as a high-performance supercapacitor electrode, *Electrochimica Acta* 192: 110–119, 10.1016/j.electacta.2016.01.140

Tripathi, V.K., Srivastava, M., Dwivedi, J., Gupta, R.K., Jangir, L.K., and Tripathi, K.M. 2024. Biomass-based graphene aerogel for the removal of emerging pollutants from wastewater. *Reaction Chemistry & Engineering* 9. 10.1039/D3RE00526G.

Uram, K., Kurańska, M., Andrzejewski, J., and Prociak, A. 2021. Rigid polyurethane foams modified with biochar. *Materials* 14(19): 5616.

U.S. Securities and Exchange Commission. 2024. Form 10-K VERDE RESOURCES, INC. For: Jun 30. http://archive.fast-edgar.com/20241016/AA22U22CZ2227JZ2222C2Z4CQOIA5C22I262

Väisänen, T., Haapala, A., Lappalainen, R., and Tomppo, L. 2026. Utilization of agricultural and forest industry waste and residues in natural fiber-polymer composites: A review. *Waste Management* 54: 62–73.

Vaughn, S.F., Winkler-Moser, J.K., Berhow, M.A., Byars, J.A., Liu, S.X., Jackson, M.A. et al. 2020. An odor-reducing, low-dust-forming, clumping cat litter produced from Eastern red cedar (Juniperus virginiana L.) wood fibers and biochar. *Industrial Crops and Products* 147: 112224.

Vaughn, S.F., Liu, S.X., Berhow, M.A., Winkler-Moser, J.K., Peterson, S.C., and Selling, G.W. 2023. Production of an odor-reducing, low-dust, clumping cat litter from soybean hulls and soybean hull biochar. *Bioresource Technology Reports* 21: 101317.

Vogler, M.R., 2023. *The Optimization of Biochar for the Removal of Phosphorus from Water*. Master's thesis, University of Idaho.

Wang, H., Liu, Q., Li, T., et al. 2021. Enhanced electrical conductivity and dielectric properties of thermally modified biochar for electromagnetic shielding applications. *Journal of Materials Research* 36(10): 1782–1791.

Wang, Y., Xie, X., Chen, Y., Niu, W., Li, Y., and Yu, J. 2022. Flexible, wearable, and biodegradable electroluminescent devices based on biochar/polymer composites. *ACS Applied Materials & Interfaces* 14(7): 9259–9270.

Wang, Y., Munir, T., Wu, X., Huang, Y., and Li, B., 2025. Phosphorus recovery and reuse: Innovating with biochar in the circular economy. *Science of The Total Environment*, 973: 179143.

Wang, Z., Li, W., Lu, Z., and Xie, Y. 2018. Experimental study on seismic behavior of reinforced concrete columns confined with hybrid fiber-reinforced polymer jackets. *Composite Structures* 184: 850–862.

Werner, S., Kätzl, K., Wichern, M., Buerkert, A., Steiner, C., and Marschner, B. 2018. Agronomic benefits of biochar as a soil amendment after its use as waste water filtration medium. *Environmental Pollution* 233: 561–568.

Wu, G., Bian, S., Wang, J., Wu, F., Sun, W., and Pan, Y. 2024. Construction of biochar-modified multi-morphologies Fe_2O_3 for efficient boosting photo-Fenton organic antibiotic pollutants degradation: Mechanism and toxicity evaluation in degradation pathways. *Industrial Crops and Products* 222: 119726.

Xu, B., Ahmed, M.B., Zhou, J.L., and Altaee, A. 2020. Visible and UV photocatalysis of aqueous perfluorooctanoic acid by TiO_2 and peroxymonosulfate: Process kinetics and mechanistic insights. *Chemosphere* 243: 125366.

Yadav, S., Ibrar, I., Al-Juboori, R.A., Singh, L., Ganbat, N., and Kazwini, T. 2022. Updated review on emerging technologies for PFAS contaminated water treatment. *Chemical Engineering Research and Design* 182: 667–700.

Yuan, X., Gao, X., Yang, J., Shen, X., Li, Z., and You, H. 2021. A sustainable and multifunctional biochar ink derived from fruit peels for flexible printed electronics. *Advanced Functional Materials* 31(52): 2105871.

Zhang, Y., He, M., Wang, L., Yan, J., Ma, B., Zhu, X. et al. 2022. Biochar as construction materials for achieving carbon neutrality. *Biochar* 4(1): 59.

Zhang, Z., Jiang, S., Chen, H., Qi, H., Chen, Y., Chen, Y. et al. 2021. Efficient solar-driven water purification based on biochar with multilevel pore bundle structure for preparation of drinking water. *Foods* 10(12): 3087.

Zhuo, Q., Liang, Y., Hu, Y., Shi, M., Zhao, C., and Zhang, S. 2023. Applications of biochar in medical and related environmental fields: Current status and future perspectives. *Carbon Research* 2(1): 32.

CHAPTER
5

Biochar's Climate Functions: Crafting a Potent Product for the Carbon Market

Arthur Gross* **and Bruno Glaser**

Martin Luther University Halle-Wittenberg, Institute of Agricultural and Nutritional Sciences, Soil Biogeochemistry, Germany

1. Introduction

With the growing urgency to mitigate climate change, carbon dioxide removal (CDR) has become a critical component of global climate strategies. Among various CDR technologies, biochar has gained prominence due to its dual role in carbon sequestration and soil enhancement. Biochar, a carbon-rich material derived from the pyrolysis of biomass, has been recognized for its ability to stabilize soil organic carbon (SOC) stocks, improve soil fertility, and contribute to long-term carbon storage.

Organic soil amendments (OSA) such as manure, compost, and crop residues have been used to improve fertility and structure of cultivated soils (Garbowski et al. 2023). Due to their organic matter content, these amendments also contribute to SOC accumulation. However, their effectiveness is often limited by rapid microbial decomposition, requiring periodic reapplications to sustain their soil benefits and long-term SOC storage. In contrast, biochar consists of highly aromatic carbon compounds, which exhibit strong resistance to microbial degradation, allowing it to persist in soils for centuries (Wang et al. 2016). This long residence time makes biochar a viable tool for active CDR, as it effectively removes CO_2 from the atmosphere and stores it in soil systems.

The increasing interest in carbon credit markets and policy-driven climate initiatives has positioned biochar as a key player in CDR strategies. The European Union's Carbon Removal Certification Framework (CRCF) aims to

*Corresponding author: arthur.gross@landw.uni-halle.de

DOI: 10.1201/9781003467946-7

establish regulatory structures for carbon sequestration methods, also affecting the integration of biochar into carbon markets. Given its high carbon retention efficiency and co-benefits for soil health, biochar is emerging as a cost-effective and scalable solution for achieving net-zero emissions targets.

However, biochar's long-term carbon storage is influenced by multiple factors, including soil type, biochar properties, and environmental conditions. Understanding the dissipation pathways of biochar—such as mineralization, metabolization, and transport mechanisms—is crucial for optimizing its carbon sequestration potential. Additionally, the economic feasibility of biochar in carbon trading systems depends on overcoming challenges such as biomass availability, production scalability, and certification standardization.

This chapter aims to: (i) evaluate biochar's role in SOC sequestration and CDR compared to other organic soil amendments, (ii) analyze biochar dissipation pathways and their implications for long-term carbon storage, (iii) assess biochar's economic and regulatory potential in carbon markets, including its alignment with the CRCF, and (iv) identify challenges and future perspectives for scaling biochar applications as a reliable climate change mitigation strategy. By synthesizing recent research and meta-analyses, this chapter provides a comprehensive overview of biochar's potential as a sustainable, long-term CDR solution, contributing to global climate targets and discussing its value in the carbon market.

2. Role of Organic Soil Amendments for Carbon Sequestration and Carbon Dioxide Removal

Since ancient times, humans have used different forms of OSA (Van Zwieten 2018), the most common forms being straw, slurries, manures and compost. More modern forms are biogas digestates, sewage sludges, biosolids and biochar. These amendments not just contain nutrients but also varying amounts of C-rich organic compounds. The application of OSA, therefore, leads to an increase in SOC stocks (Alvarenga et al. 2020). If this additional organic carbon can be stabilized in soils, the approach can be considered a viable method for active, long-term CDR from the atmosphere.

Different types of OSA contain diverse carbon compounds, each with varying stability and mean residence time (MRT) in soil after application. Solid manures and compost usually show higher dry matter contents and a higher C/N ratio, leading to longer soil persistence ranging between several weeks and months. Consequently, these amendments need to be applied regularly, e.g. at the beginning of a new growing season to contribute to soil fertility and productivity, and an SOC stock increase in longer terms (Gross and Glaser 2021; Joona et al. 2024). In contrast, biochar amendments contain highly aromatic C compounds and only little amount of N (most of it being polycyclic and not available for plants or microorganisms) (Singh et al. 2014)), and are, therefore, resistant against microbial decomposition. When applied to soil, biochar's MRT is estimated to be

556±483 years (Wang et al. 2016), which underlines its high SOC sequestration potential. The magnitude of SOC stock increase was found to vary strongly between the different OSA types, as a range of different global meta-analysis suggested (Table 1).

Table 1. Compilation of recent meta-analysis results of organic soil amendments effects on soil organic carbon stocks. CI = Confidence interval, n = Number of observations

Organic soil amendment	SOC stock increase [95% CI]	Long-term SOC stock increase	Note	Reference
Compost	29% (25–33%)	30%	Long-term effect estimated across all application frequencies over 25 years	Bai et al. 2023
Slurry	15% (10–15%)	16%		
Sewage sludge	80% (41–119%)	n.s.		
Manure	26% (24–28%)	36%		
Biosolids	72% (57–87%)	38%	Long-term effect estimated across all application frequencies over 11–30 years; single application: n.s	Snyder 2021
Biochar	29% (26–33%)	49%	Single application effect after 10 years	Gross et al. 2021

The addition of various forms of OSA to soil are a direct input of organic C compounds to soil. Therefore, elevated SOC stocks after the application of these amendments are expected. However, to understand if these amendments lead to SOC sequestration and, therefore, to enhanced SOC storage in the long term, one must distinguish the types and properties, especially the stability of organic C compounds amended to soil.

Adding conventional OSA such as manure to soil increases SOC stocks on a global mean by 10.7 Mg ha^{-1}, thus close to the mean SOC increase after biochar application (Figure 1). Combining this with the fact that manure nitrogen application to cropland accounts for less than one-fifth of the total manure nitrogen production, there seems to be large potential for CDR via manure-driven SOC sequestration. However, organic matter in manure is rapidly degraded due to its high nitrogen content or its low carbon-to-nitrogen ratio and due to the dominance of labile organic compounds (Gross and Glaser 2021). Therefore, even continuous seasonal or annual applications of manure did not lead to significant higher SOC stocks between different duration classes (Figure 1). Biochar in contrast, even applied once, continuously increased the SOC stock with increasing time up to 10 years (Figure 1). This observation, on the one hand, underlines biochar's persistence and on the other hand the potential for SOC stabilization. If this trend

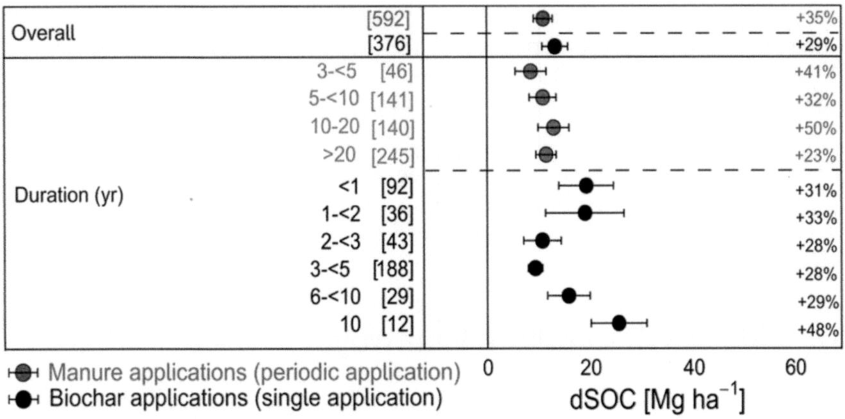

Figure 1. Meta-analysis results of manure and biochar application on SOC level changes, given as a forest plot. Presented is the mean difference and relative change of soil organic carbon stocks (SOC) after manure and biochar application (dSOC) differentiated by different duration since application. Number in brackets represent the number of included treatments. Points within the range represent the mean dSOC and the line within the 95% confidence interval represents the range of the effect size. The result can be significantly different at $p < 0.05$ from each other if the 95% confidence intervals did not overlap. Data obtained from Gross and Glaser (2021) and Gross et al. (2021).

continues in longer time frames >10 years remains unknown, due to limited observations on decadal scales.

If a certain biomass should be used for CDR is also a question of prioritization. Available manure should first and foremost be used for fertilization and thus for food production. While selecting suitable areas for manure application, the potential to enhance food production should be prioritized over CDR, since available area is limited. Additionally, practitioners should be aware of nutrient thresholds in soil while selecting areas for application of manure. Pyrolyzing manure or sewage sludge can serve as a solution, as local application may be restricted due to local thresholds and transporting it elsewhere would generate additional emissions. However, because pyrolysis leads to the loss of valuable nutrients, especially N, its ecological viability should be evaluated holistically—for example, through life cycle assessments conducted beforehand, to avoid negative trade-off effects. A more sustainable approach would be to co-compost manure or sewage sludge together with biochar. Thereby, nitrogen leaching can be avoided (Kammann et al. 2015), without negatively affecting biochars SOC sequestration potential (Apostolović and Gross et al. 2024).

Generally, only biomass that does not fulfill another important purpose and is nutrient-poor should be considered for pyrolysis, especially woody residues. Current estimates quantify the available biomass to be ~152 Mt in Germany (DBFZ 2020) and ~580 Mt in the EU (Tripathi et al. 2020). Using the assumption of a pyrolysis yield of 30% at 600 °C (Altıkat et al. 2024), this would be equal to a biochar mass of 174 Mt, and each t saves ~3 t CO_2-eq, that would be the total equivalent of 522 Mt CO_2-eq. Even with less optimistic estimates and achieving only a portion of that, this number underlines the large potential to save emissions using biochar, which we urgently need to in order to meet the goal of the EU "Green Deal" of saving 310 Mt of CO_2-eq. Since pyrolysis production capacities are exponentially growing with a compound annual growth rate of 91% (IBI 2024), the potential for widespread implementation of biochar systems and thus saving these emissions exists.

3. Biochar Dissipation Pathways

Once applied to soil, biochar does not remain rigidly in place but reacts with the environment including soil flora, fauna and minerals as well as plants and water. This involves various dissipation processes including mineralization, metabolization and vertical and lateral translocation of biochar. Even though there are few individual studies describing one of these mentioned biochar dissipation pathways, describing biochar dissipation dynamics holistically with respect to all relevant dissipation processes, have been proven challenging. Moreover, the question of how these biochar dissipation dynamics influence the ability to sequester SOC and thus, biochar's CDR potential, remains largely unknown, since there is a general lack of experimental data on biochar SOC sequestration effects in the long-term and under different real field conditions.

Soil texture is one of the most important influencing factors of biochar dissipation (Berhane et al. 2020; Gross et al. 2024; Han et al. 2016). Figure 2 adopted from Gross et al. (2024) summarizes the key mechanisms determining biochar dissipation in a fine (Figure 2a) and coarse textured soil (Figure 2b).

Soils with finer textures contain higher levels of clay minerals and iron oxides, which contribute to the protection of SOC from enzymatic breakdown and turnover (Nannipieri et al. 2018) and increases SOC stability in the form of physical protection within soil aggregates. Sandy soils provide less of this protection, making SOC more vulnerable to oxidation and decomposition (Gross et al. 2021), along with losses of SOC through leaching and runoff (Yang et al. 2019). Notably, an increase in native SOC following biochar addition has been observed, a phenomenon often described as the "negative priming effect" (Blanco-Canqui et al. 2020; Guo et al. 2024). This effect tends to persist in conditions of reduced tillage and on loamy soils (Blanco-Canqui et al. 2020; Gross et al. 2024).

Other major factors influencing the dissipation of biochar are its properties (Figure 2). The key property influencing biochar stability is the H/C ratio, an indicator of biochar aromaticity (Budai et al. 2013; Schimmelpfennig and Glaser 2012). The H/C ratio is an indicator for the aromaticity of biochar. Compared to uncharred biomass, which typically has higher H/C ratio, biochar with low H/C ratio is generally expected to be more stable over the long term (Budai et al.

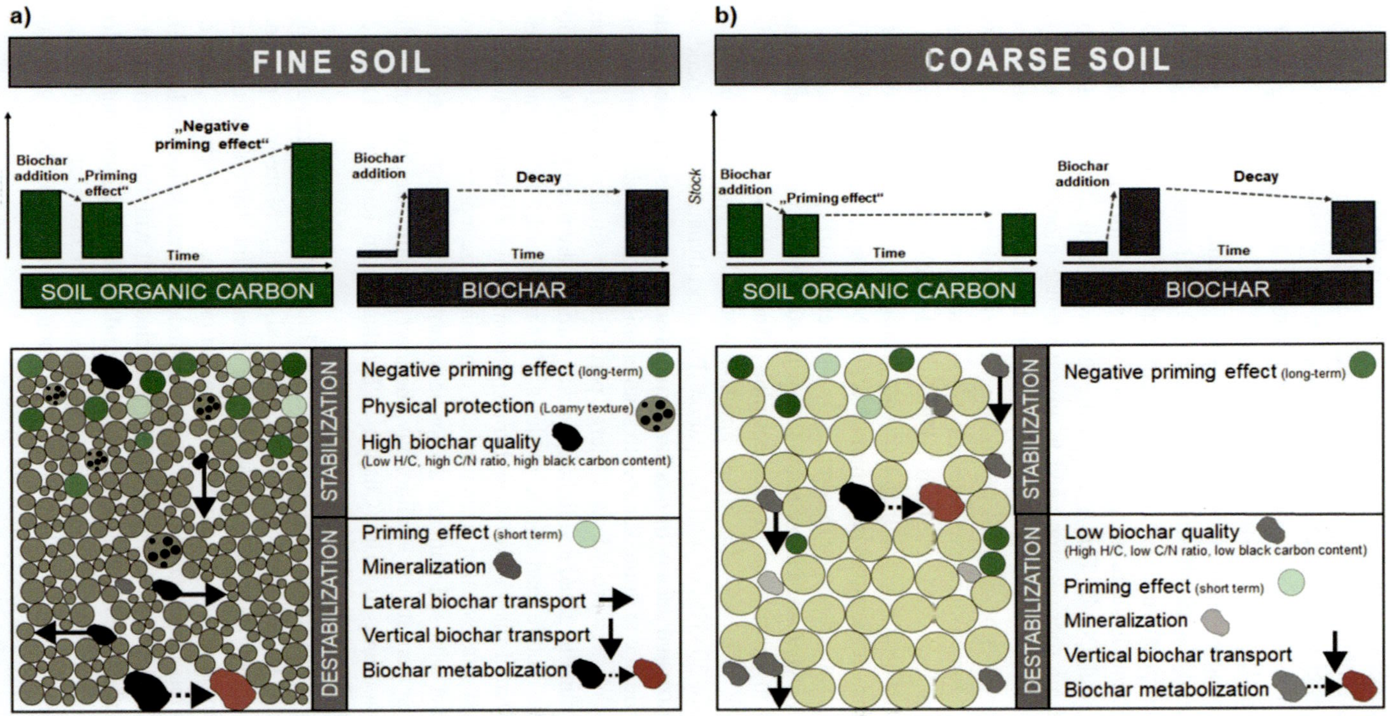

Figure 2. Overview of soil organic carbon (SOC) and biochar behaviour in fine (a), and coarse soil (b). The upper part shows (SOC) and biochar stock dynamics over time and the responsible processes. The bottom part shows mechanisms leading to stabilization and destabilization of SOC and biochar. Inspired by Gross et al. (2024).

2013). The biochar in the loamy soil (Figure 2a) was made of lignin-rich material and exhibits a higher C/N ratio than the biochar used in the sandy soil (Figure 2b). The carbon and nitrogen content in biochar, along with the C/N ratio, are also critical for determining its stability and potential for contributing to SOC accumulation (Liu et al. 2016).

Biochar metabolization is another possible biochar dissipation pathway. Biomarker analysis has shown that from biochar's total mass loss of ~0.3% per year, the incorporation of biochar into microorganisms is responsible for up to 1.5% to 2.6% and is therefore a negligible carbon source for soil microorganisms (Kuzyakov et al. 2014). This is due to the relatively high amount of activation energy that microorganisms need to invest, to use biochar compounds for metabolization (Lehmann et al. 2024). Even if metabolization accounts for a negligible proportion of total dissipation in terms of quantity, this pathway is important for an accurate determination of total biochar mineralization – particularly to have a reliable basis for the use of biochar for carbon credits.

Biochar transport is the major driving factor of dissipation and occurs in two directions. Vertical transport, mainly driven by tillage, rainfall, bioturbation and soil physical properties (Major 2010; Obia et al. 2017), occurs in rates up to 30 mm per year and can account for up to 40–70% of the applied biochar, as observed in agricultural experiment plots (Rumpel 2024). Lateral transport is mainly driven by erosion, tillage and soil water flow and was found to be responsible for 20–53% of total dissipation. Transport dynamics also depends on the particle size and density of biochar particles and aggregates. Larger particles are less mobile due to limited access to smaller soil pores. Physical disintegration of biochar leads to a breakdown of large particles into smaller sizes (Lehmann et al. 2024). Smaller sized and lighter biochar particles are more prone to erosion and are more mobile than coarse particles. It is noteworthy that in most cases biochars physical disintegration and transport only have a minor influence on mineralization rates. Smaller sized particles can mineralize more rapidly; however, they also react with soil minerals and/or be incorporated into aggregates, which increases their persistence. Additionally, the inert aromatic backbone of biochar can persist thousands of years (Chapter 1 "Biochar: From Terra Preta Discovery to Modern Industry").

Due to its low mean decay rate of ~0.3%, biochar has an enormous carbon sink potential (Wang et al. 2016; Woolf et al. 2021). Moreover, previous estimations of biochar's longevity were often based on incubation studies, which used exponential models to calculate mineralization rates. In most cases, they did not integrate biochars inert pool and, therefore, underestimated biochars long-term stability (Li et al. 2024). Although biochar particles react with the environment in many ways, their use for carbon sequestration and CDR is highly recommended. However, dissipation of biochar, especially transport dynamics largely influences its CDR potential, since CDR measures for carbon credits should be targeted and accountable, also years after soil addition.

3. Biochar Potential for CDR Credits

Efficient CDR technologies are urgently needed. The European Union (EU) is pursuing CO_2 neutrality by 2050 as part of the "Green-Deal". To achieve this ambitious goal, the EU Commission proposed a net reduction in emission of GHG of 55% by 2030 compared to 1990 emission levels and net-neutrality by 2050. Net-neutrality means full compensation of all residual emissions beyond 2050. To realize this, it is required to increase natural sinks, e.g. through restoration of peatlands, afforestation, and agroforestry. However, these measures have limited potential, are not permanent and require areas which potentially stay in conflict with other forms of usage, e.g. food and energy production, settlements, and natural habitats.

The use of biochar for carbon removal (BCR) is more promising because of its reliability and minimal risk of being used for green washing. Furthermore, biochar production via pyrolysis is relatively easy to implement and can be realized decentrally. The two major limits are available biomass and its relatively high investment costs (Gross et al. 2023). By introducing biochar into carbon credit markets, farmers have an additional financial motivation to use biochar to amend their soils, besides various positive effects on soil health in the long-term (Apostolović and Gross et al. 2024). Considering the growing production capacity (IBI 2024) for biochar and the increasing CO_2 price (Gross et al. 2023), biochar should become affordable and ready-to-use on a broad scale. Depending on the feedstock and production method, biochar typically contains over 80% carbon, meaning one ton of biochar equates to roughly three tons of CO_2 equivalent. The current price for a ton of CO_2 in the European Emissions Trading system is €70 (Statista 2024), meaning that one ton of biochar represents a value of approximately €210. By 2030, however, the CO_2 price could already be well over €250 (Endres 2023), which would make the use of biochar for CO_2 compensation much more profitable. Furthermore, BCR is among the cheapest CDR measures which are currently being sold (CDR.fyi 2024).

Recently in July, 2024, the EU commission adopted the provisional agreement from 2022 on the Carbon Removals and Carbon Farming (CRCF) regulation for an EU-wide voluntary CDR certification framework which aims to increase the deployment speed of CDR, provide evaluation criteria for the quality of emission reduction measures, and to counter greenwashing and build trust (Fallasch et al. 2024). Such an EU-wide regulation could provide the legal foundation for an broad-scaled carbon credit program, business opportunities for landowners and the carbon credit industry, and could act as a role model for different world regions.

Currently, the majority of companies and projects offering BCR are concentrated in Europe, with only a small number operating in South America and Africa (CDR.fyi 2024). However, the greatest potential for photosynthesis-driven CDR lies in tropical and subtropical regions, making agricultural enterprises in these regions – including smallholder farms key players in biochar production and the CDR certification system (Carbon Standards International 2022).

One limiting factor is that the standards for well-established certifications, such as the "European Biochar Certificate" (EBC) and the forthcoming "Global Biochar Carbon Sink Standard", are tailored to European conditions and can be expensive to fulfill for practitioners in the Global South, e.g., laboratory analyses to ensure high biochar quality are required.

However, biochar of high-quality and permanence can also be produced at low cost using Kon-Tiki reactors. To make biochar production in resource-limited settings more feasible, the Ithaka Institute for Carbon Strategies has developed the "Global Artisan C-Sink Certification System" (Carbon Standards International 2022). This system provides specific guidelines to support low-tech biochar producers in obtaining carbon sink certificates, enabling them to participate in the global BCR market. By implementing such accessible frameworks, the huge CDR potential of the Global South can be increasingly unlocked.

These recent progresses on political, legal and economic scale provide good chances for wide-scaled use of BCR, even beyond the EU borders. At the same time, there is still huge demand for further research. Long-term field observations including dissipation pathways are still rare but are urgently needed to provide the empirical foundation to integrate biochar into the carbon market. Unused biomasses, suitable for pyrolysis, must be identified based on holistic assessment methods, to avoid potential negative trade-offs. Last but not least, certification methods for carbon sinks have to be unified, e.g. they should harmonize with IPCC guidelines, to create a reliable work-basis, and incentives for companies, landowners and governments.

4. Conclusion and Recommendations

Biochar has proven to be a highly effective CDR technology, offering both long-term SOC sequestration and additional agricultural and environmental benefits. Unlike conventional OSA such as manure and compost, biochar exhibits exceptional stability due to its high aromatic carbon content, making it a viable long-term carbon sink. The ability of biochar to persist in soils for decades (probably even centuries), along with its negative priming effect on SOC, reinforces its role as a scalable CDR strategy.

Meta-analyses confirm that biochar enhances SOC stocks more durable over time compared to other OSA, particularly in loamy soils, where physical stabilization further extends its persistence. However, biochar's dissipation pathways, including mineralization, metabolization, and lateral and vertical transport, have implications for its long-term sequestration potential. While biochar degradation rates remain low (~0.3% per year), continued research under real field conditions is needed to assess its stability over decadal timescales.

Biochar is increasingly being integrated into carbon credit markets, with the CRCF providing regulatory clarity for its certification. The growing CO_2 price and the expansion of biochar production capacities further position biochar as a cost-effective climate solution. However, to unlock its full potential, challenges related to certification standardization, biomass sourcing, and long-term field data must be addressed.

To optimize biochar's CDR potential and market adoption, the following actions are recommended:

1. Expand long-term field experiments
   - Conduct or re-assess long-term field experiments to analyze biochar's persistence, stabilization mechanisms, and interaction with native SOC
   - Improve understanding of biochar dissipation pathways, particularly its behavior in differently textured soil, under different climate and tillage, to refine sequestration estimates
2. Enhance biochar certification and carbon credit systems
   - Align biochar certification methodologies with international carbon accounting frameworks (e.g., IPCC guidelines)
   - Promote the development of low-cost certification models (e.g., Global Artisan C-Sink Certification) to support small-scale biochar producers in the Global South
   - Improve MRV (Measurement, Reporting, and Verification) frameworks to standardize biochar-based carbon credits in international markets
3. Prioritize biomass utilization for pyrolysis
   - Use nutrient-poor residues (e.g., woody biomass, agricultural by-products) for pyrolysis to maximize carbon sequestration without compromising soil fertility
   - Avoid pyrolyzing high-nutrient organic matter (e.g., manure) unless paired with co-composting strategies to prevent nutrient loss
   - Pyrolysis of sewage sludge shall be prioritized over its burning because of elimination of organic pollutants but recycling of nutrients such as N, P, Ca, Mg, K etc.
4. Optimize biochar application strategies
   - Tailor biochar application rates based on soil texture, climate, and land use, considering trade-offs between sequestration and agricultural benefits
   - Implement co-application strategies (e.g., biochar + compost or biochar + manure) to enhance nutrient retention while preserving carbon stability
5. Scale biochar adoption through policy and market incentives
   - Introduce financial incentives and subsidies for farmers and industries using biochar as a soil amendment
   - Establish government-backed biochar deployment programs within climate action plans (e.g., EU Green Deal, National CDR Strategies)
   - Encourage corporate investments in biochar carbon credits, similar to Google's 100,000-ton CO_2 removal purchase from biochar projects in 2025

Summarizing, biochar presents a highly scalable, cost-effective, and scientifically validated solution for carbon sequestration and climate change mitigation. However, unlocking its full potential requires policy support, regulatory standardization, and continued research. By integrating biochar into global CDR markets, sustainable agriculture, and soil health management, it can serve as a cornerstone technology for achieving net-zero emissions while fostering resilient ecosystems and carbon-neutral economies.

References

Altıkat, A., Alma, M.H., Altıkat, A., Bilgili, M.E., and Altıkat, S. 2024. A comprehensive study of biochar yield and quality concerning pyrolysis conditions: A multifaceted approach. *Sustainability* 16: 937.

Alvarenga, P., Carneiro, J.P., Fangueiro, D., Cordovil, C.M.d.S., and Bernal, M.P. 2020. Managing organic amendments in agroecosystems to enhance soil carbon storage and mitigate climate change. *Climate Change and Soil Interactions* 89–141.

Apostolović, T., Gross, A., Rodríguez, Á.F.G., de la Rosa, J.M., Glaser, B., Knicker, H. et al. 2024. Impact of biochar aging on soil physicochemical properties. *Agronomy* 14: 3007.

Berhane, M., Xu, M., Liang, Z., Shi, J., Wei, G., and Tian, X. 2020. Effects of long-term straw return on soil organic carbon storage and sequestration rate in North China upland crops: A meta-analysis. *Global Change Biology* 26: 2686–2701.

Blanco-Canqui, H., Laird, D.A., Heaton, E.A., Rathke, S., and Acharya, B.S. 2020. Soil carbon increased by twice the amount of biochar carbon applied after 6 years: Field evidence of negative priming. *GCB Bioenergy* 12: 240–251.

Budai, A., Zimmerman, A.R., Cowie, A.L., Webber, J.B.W., Singh, B.P., Glaser, B. et al. 2013. Biochar carbon stability test method: An assessment of methods to determine biochar carbon stability. *International Biochar Initiative*. https://biochar-international.org/wp-content/uploads/2018/04/IBI_Report_Biochar_Stability_Test_Method_Final.pdf

Carbon Standards International. 2022. *Guidelines for Carbon Sink Certification of Artisan Biochar Production* – Version 1.0.

CDR.fyi. 2024. *Global CDR Leaderboard*. Retrieved from www.cdr.fyi.

DBFZ. 2020. *DBFZ Ressourcendatenbank*.

Endres, L. 2023. Verteilungswirkung der CO_2-Bepreisung in den Sektoren Verkehr und Wärme mit Pro-Kopf Klimageld. *IMK Policy Brief* 161.

Fallasch, F., Böttcher, D.H., Schneider, D.L., Herrmann, H., Jörß, W., Siemons, A. et al. 2024. The EU Carbon Removal Certification Framework: Options for using certified removal units and funding mitigation activities. *Climate Change* 44. https://www.umweltbundesamt.de/sites/default/files/medien/11850/publikationen/44_2024_cc_the_eu_carbon_removal_certification_framework.pdf

Garbowski, T., Bar-Michalczyk, D., Charazińska, S., Grabowska-Polanowska, B., Kowalczyk, A., and Lochyński, P. 2023. An overview of natural soil amendments in agriculture. *Soil and Tillage Research* 225: 105462. https://doi.org/10.1016/j.still.2022.105462

Google. 2025. URL: https://blog.google/feed/were-announcing-our-first-partnerships-to-scale-biochar-for-co2-removal/ (accessed 2.18.25).

Gross, A., Bromm, T., and Glaser, B. 2021. Soil organic carbon sequestration after biochar application: A global meta-analysis. *Agronomy* 11: 2474.

Gross, A., and Glaser, B. 2021. Meta-analysis on how manure application changes soil organic carbon storage. *Scientific Reports* 11: 5516.
Gross, A., Möller, A., and Glaser, B. 2023. *Pflanzenkohle-Systeme zur Kohlenstoffspeicherung in Böden*. Universität Hamburg.
Gross, A., Bromm, T., Polifka, S., Fischer, D., and Glaser, B. 2024. Long-term biochar and soil organic carbon stability – Evidence from field experiments in Germany. *Science of The Total Environment* 954: 176340.
Guo, F., Wang, C., Wang, S., Zhao, X., Li, G., and Sun, Z. 2024. The native SOC increase in woodland and lawn soil amended with biochar surpassed greenhouse—A seven-year field trial. *Science of The Total Environment* 907: 167924.
Han, P., Zhang, W., Wang, G., Sun, W., and Huang, Y. 2016. Changes in soil organic carbon in croplands subjected to fertilizer management: A global meta-analysis. *Scientific Reports* 6: 27199.
IBI. 2024. *2023 Global Biochar Market Report*. International Biochar Initiative.
Joona, J., Liski, E., and Kahiluoto, H. 2024. Manure increases soil organic carbon most when allocated to annual cropping. *Catena* 238: 107844.
Kammann, C.I., Schmidt, H.-P., Messerschmidt, N., Linsel, S., Steffens, D., Muller, C. et al. 2015. Plant growth improvement mediated by nitrate capture in co-composted biochar. *Scientific Reports* 5: 11080.
Kuzyakov, Y., Bogomolova, I., and Glaser, B. 2014. Biochar stability in soil: Decomposition during eight years and transformation as assessed by compound-specific ^{14}C analysis. *Soil Biology and Biochemistry* 70: 229–236.
Lehmann, J., Abiven, S., Azzi, E., Fang, Y., Singh, B.P., Sohi, S. et al. 2024. Persistence of biochar. *Biochar for Environmental Management* 277–311.
Li, H., Azzi, E.S., Sundberg, C., Karltun, E., and Cederlund, H. 2024. Can inert pool models improve predictions of biochar long-term persistence in soils? *Geoderma* 452: 117093.
Liu, S., Zhang, Y., Zong, Y., Hu, Z., Wu, S., Zhou, J. et al. 2016. Response of soil carbon dioxide fluxes, soil organic carbon and microbial biomass carbon to biochar amendment: A meta-analysis. *GCB Bioenergy* 8: 392–406.
Major, J., Lehmann, J., Rondon, M., and Goodale, C. 2010. Fate of soil-applied black carbon: Downward migration, leaching, and soil respiration. *Global Change Biology* 16: 1366–1379.
Nannipieri, P., Trasar-Cepeda, C., and Dick, R.P. 2018. Soil enzyme activity: A brief history and biochemistry as a basis for appropriate interpretations and meta-analysis. *Biology and Fertility of Soils* 54: 11–19.
Obia, A., Børresen, T., Martinsen, V., Cornelissen, G., and Mulder, J. 2017. Vertical and lateral transport of biochar in light-textured tropical soils. *Soil and Tillage Research* 165: 34–40.
Rumpel, C. 2024. Biochar transport in terrestrial ecosystems. *Biochar for Environmental Management* 313–330.
Schimmelpfennig, S., and Glaser, B. 2012. One step forward toward characterization: Some important material properties to distinguish biochars. *Journal of Environmental Quality* 41: 1001–1013.
Singh, B., Fang, Y., Cowie, B.C.C., and Thomsen, L. 2014. NEXAFS and XPS characterisation of carbon functional groups of fresh and aged biochars. *Organic Geochemistry* 77: 1–10.
Statista. 2024. *EU-Emissionshandel: Entwicklung der Preise 2024*. Retrieved from Statista.
Wang, J., Xiong, Z., and Kuzyakov, Y. 2016. Biochar stability in soil: Meta-analysis of decomposition and priming effects. *GCB Bioenergy* 8: 512–523.

CHAPTER
6

Developing Biochar Carbon Credit Projects: A Simplified Guide

Suzanne Allaire
GECA Environnement, 1516 Ch. Du Club-Nautique, Lac-Sergent, QC. Canada

1. Introduction

Biochar is a black, carbon-rich material made when plant or organic waste is heated without oxygen to temperature varying from more than 300 °C up to 900 °C. This process is called pyrolysis. Because there is no oxygen, the carbon in the biomass does not burn but becomes stable and solid. This makes biochar very helpful in the fight against climate change.

Indeed, when biochar is made and used correctly, the carbon from plant is locked into a stable form, called recalcitrant. The recalcitrant carbon cannot be degraded by the microorganisms. Once in soil or in other environments, it stays there for hundreds of years. This mimic a similar process naturally occurring in nature like forest fires that generate charcoal when the wood is not completely consumed followed by soil developing or sliding above it and locking the charcoal. This old charcoal is used to date fire episode thousands of years back. The earth has similar other locking systems, which created, a long time ago, the coal, the natural gas, and the petroleum. When we extract these fossil fuels from the ground, we put the carbon back to the atmosphere the nature had locked. Biochar is used to lock back carbon to the ground or other materials (Figure 1).

The recognition of the good action to lock carbon in the ground or in other materials allows access to financial support, called "carbon credits". A carbon credit is a certificate that shows that one tonne of CO_2 has been taken out of the air or avoided. Of course, this recognition must be proven and go through audit

Email: suzanne.allaire@gecaenviro.com

DOI: 10.1201/9781003467946-8

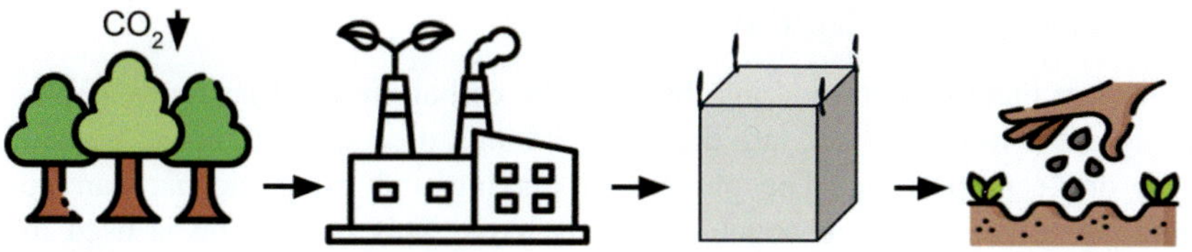

Figure 1. Fixing atmospheric carbon through biochar in soil.

and certification by international recognized organisations. These carbon credits can be sold to companies or individuals who has quantified their carbon mission and want to wish to compensate for it, making it a financial opportunity for those locking this carbon. In other words, the entities that lock carbon through biochar in soil, in underground or in construction materials are applauded by others that emit, through the sale of the carbon credits from the lockers to the emitters.

This book chapter is for anyone who is curious about starting or investing in biochar carbon credit projects. Whether you're a business owner, an investor, a policymaker, or just someone interested in climate solutions, this chapter explains the key ideas in biochar carbon credits in easy language. No need to have a background in science or carbon markets to benefit from this chapter.

2. Understanding Biochar Carbon Credits

What are Carbon Credits?

A carbon credit is a kind of "certificate" that shows one tonne of CO_2 has been removed or avoided from the atmosphere, the removal does not have to be directly from the air. It is accepted than the plants extract the CO_2 from the atmosphere during their growth to store it in their structure. This is usually this carbon that we transform into biochar to fix it. One credit equals one tonne of carbon dioxide that will not contribute to global warming because it is no longer in the atmosphere. These credits are sold in two kinds of markets:

- **Voluntary market:** People or companies buy credits because they want to, not because a law tells them to. These buyers often want to show their commitment to the environment or meet their sustainability goals.
- **Compliance market:** Governments require companies to buy credits if they go over a set emission limit. These are legally regulated.
- *For now, biochar carbon credits are only used in the voluntary market.* This means that project developers can sell their credits, but they must find buyers who are not legally required to buy them.

How does Biochar Qualify?

When plant material is turned into biochar, the carbon inside it becomes stable. This means it won't go back into the air for a long time, unlike compost, burned wood, or decomposition. The process of making and using biochar can be monitored to measure how much carbon is fixed. This monitoring is used to create carbon credits.

To qualify for carbon credits, a biochar project must follow strict rules defined by carbon standards such as Puro.Earth, European Biochar Certificate (EBC), and others. These standards require proof that the carbon in the biochar is permanent, meaning it will stay locked in the soil or another storage location for 100 + years.

Biochar must be tested in a laboratory to verify its fixed carbon content, H/C ratio (hydrogen to carbon), and O/C ratio (oxygen to carbon). A low H/C ratio (typically under 0.7) means the biochar is stable and won't break down nor be digested easily. These results are essential inputs for the Life Cycle Assessment (LCA), which measures the sequestration and the emissions, including emissions from transport and processing, of carbon to make a carbon budget, meaning the total carbon removal.

The project must also show how it avoids emissions that would have occurred if the biomass was left to rot or burn. This is called the baseline scenario. All data and test results must be recorded, reviewed, and verified by an independent third party through auditing party before any carbon credits can be issued. They must be issued before being sold (Figure 2).

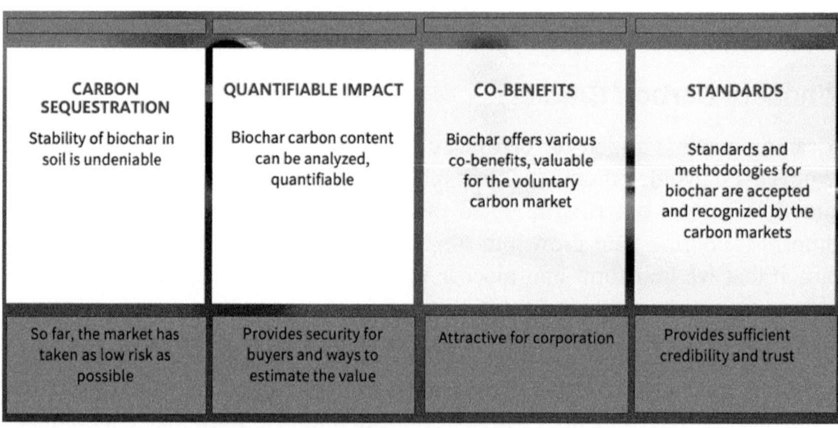

Figure 2. How biochar fits into the carbon market.

The Science Behind It

The more stable the carbon, the longer it stays out of the atmosphere (Woolf et al. 2021) This makes biochar a powerful tool for long-term carbon removal. Science was needed to show the concept, how it works and how to measure it. Now that it is accepted worldwide by official entities, science is needed to measure the real impact of each project on carbon removal.

To verify this stability of the carbon in the biochar, scientists measure the hydrogen-to-carbon (H/C) and oxygen-to-carbon (O/C) molar ratios. A low H/C ratio (typically under 0.7 indicates that the biochar is chemically stable and resistant to microbial degradation. Similarly, a low O/C ratio signals that the carbon structure is aromatic and not easily broken down by natural processes. High fixed carbon content and low volatile matter are also signs of good stability and long-term carbon storage.

Certified laboratories conduct these tests, and the results are used in the project's Life Cycle Assessment (LCA). The LCA calculates the net carbon removal by considering not only the carbon stored in the biochar but also emissions from transport, processing, and any energy used. This scientific basis ensures the carbon credits are real, measurable, and verifiable.

3. Steps to Developing a Biochar Carbon Project

Step 1: Project Feasibility & Business Model

Select your Feedstock Wisely

Start by identifying the biomass you will use—options include agricultural residues, forestry by-products, agro-food wastes, sludge or other plant-based waste materials. The feedstock must be both sustainable and ethically sourced. It's essential to avoid materials that could otherwise serve high-value purposes, such as food or feed. By prioritizing true waste streams (e.g., crop residues, sawdust, pruning waste), you help reduce pressure on land use, landfill, avoid indirect emissions, and strengthen the environmental case for your project.

Choose the Appropriate Technology

The majority of biochar projects rely on pyrolysis, a thermal decomposition process in the absence of oxygen. However, other thermal conversion technologies like gasification (partial oxidation), carbonization, and torrefaction (mild pyrolysis) may also be suitable depending on your goals. Each method varies in terms of CAPEX, OPEX, carbon retention efficiency, energy recovery, and the quality of both biochar and by-products (like syngas or bio-oil). Choose the one that aligns best with your operational capacity, carbon objectives, and desired co-products.

Conduct thorough Market Research

Analyze your local and regional markets to determine the demand for both biochar (e.g., in agriculture, landscaping, construction materials) and carbon credits (compliance or voluntary markets). Make sure that the market is eligible for carbon credits. For example, most of the metallurgy applications are not as for high tech, battery, cosmetics and other markets. Understanding buyer needs, regulatory frameworks, pricing trends, and potential partners will help you craft a more resilient business strategy. A deep understanding of the market is also crucial for investor engagement, partnership building, and long-term project stability. Take a specialist to help you navigate those markets.

Build Detailed Revenue Projections and Business Models

Successful biochar projects often combine multiple income streams. Primary revenues usually come from the sale of biochar products and carbon removal credits, but you may also consider revenue from co-products (like heat, electricity or bio-oil) or public funding. Forecast both your operating costs and potential returns across various scenarios (e.g., low vs. high credit prices), and ensure your model reflects realistic timelines for credit issuance and payment. This integrated financial outlook will be essential for attracting investors and sustaining operations over time.

Step 2: Certification and Methodologies

Choose a Standard

Puro.Earth and *European Biochar Certificate (EBC)* are currently active. Other standards under development or in use include *Verra* (Verified Carbon Standard), *Climate Action Reserve (CAR)*, *Carbon Standards International (CSI)*, *Isometric, Carbon Future*, and the new *Label Bas Carbone France* (LBC France), which is the official French carbon certification framework.

Each of these standards has different methodologies to estimate the carbon being sequestered, geographic reach, and requirements, so it's important to choose one that matches your project's size, goals, and location. These standards help ensure your project meets high environmental integrity and is eligible to sell carbon credits on trusted marketplaces. These organizations provide the rules and systems to issue carbon credits.

Collect data: Data collection is one of the most critical parts of a carbon project. You must accurately track how much biochar is produced, its moisture content, fixed carbon content, and the hydrogen-to-carbon (H/C) and oxygen-to-carbon (O/C) ratios. You also need to document feedstock types and quantities, pyrolysis temperatures and durations, energy consumption, emissions from the process, emissions from transportation, packaging, and other emissions, and end uses of the biochar. Laboratory analysis is required to test the biochar for fixed carbon, ash content, volatile matter, pH, nutrient content, and heavy metals. These lab results are essential inputs for calculating the Life Cycle Assessment (LCA), which determines your project's net carbon removal. The data must be gathered consistently, with timestamps and traceability, and should be backed up with photos, lab certificates, and digital records to meet auditing and verification requirements. High-quality data ensures that your carbon removal claims can be proven and certified with confidence.

Methodologies Set the Rules

They define how to calculate carbon storage, report your results, and get verified. Each standard provides detailed methodology documents that list all the steps a project must follow to be eligible for certification. This includes how to select system boundaries for your carbon accounting, how to run a Life Cycle Assessment (LCA), and what documentation and laboratory results must be submitted. Biochar must meet specific technical criteria, including minimum fixed carbon content, maximum levels of ash or volatile matter, and stability indicators such as H/C and O/C molar ratios. Following a recognized methodology gives your project credibility and makes your credits easier to sell.

Step 3: Implementation and Scaling

Technology

Choose a reliable and clean system. The choice of technology depends heavily on local conditions, available investment, and the project's scale. Batch pyrolysis systems, which require manual loading and unloading, are more suitable for regions with lower labor costs or where small-scale, decentralized production is preferred—such as in many developing countries. These systems are typically simpler, less expensive, and easier to maintain, making them ideal for rural or remote areas. On the other hand, continuous pyrolysis systems are better suited for industrialized regions or for large-scale operations where labor is costly and a high tonnage of biochar is required. These systems automate most steps, reduce operational downtime, and can produce consistent-quality biochar at a larger volume. Choosing the right technology also affects energy efficiency, emissions, and the ability to meet certification requirements.

Location

Think about where feedstock comes from and how close you are to customers. Reducing the distance from your feedstock source to your facility lowers transportation costs and reduces greenhouse gas emissions from logistics. Similarly, minimizing the distance to your biochar users or markets is equally important. Selling biochar locally helps improve your project's environmental footprint, increases social acceptance, and can make distribution more efficient and profitable. A good location helps reduce transportation costs and emissions.

Energy and Distribution

You might use the heat produced to run your plant or sell electricity. A more sustainable approach is to reuse the syngas produced during pyrolysis. Syngas, or synthesis gas, is a mixture of gases (mainly carbon monoxide, hydrogen, and methane) that results from breaking down biomass without oxygen. Instead of venting it, this gas can be captured and burned to generate heat for the reactor itself, making the process more energy-efficient and reducing fossil fuel use.

Reusing syngas can also improve your project's carbon footprint and help meet certification requirements. This adds another income stream and improves the project's sustainability.

Step 4: MRV (Measurement, Reporting, and Verification)

MRV is the backbone of any credible carbon project. It ensures that all claimed carbon removals are real, accurately calculated, and not duplicated. MRV helps project developers track performance, prove compliance with standards, and build confidence among buyers.

Measurement

Start with detailed measurement of biochar quantities, properties (such as moisture content, fixed carbon, and volatile matter), and production process data (temperature, feedstock type, energy used, and emissions). Laboratory analysis must be conducted to determine carbon stability indicators like H/C and O/C ratios.

Reporting

All data should be carefully recorded and presented in a clear, structured format. Reports typically include operational data, biochar analysis results, and calculations of net carbon removals using Life Cycle Assessment (LCA). The LCA evaluates emissions and energy use throughout the biochar production chain, from feedstock sourcing to final application or sale.

What is Life Cycle Assessment (LCA)?

LCA is a systematic methodology for quantifying the environmental impacts of a product, process, or service across its entire life cycle. For biochar projects, this means assessing all greenhouse gas emissions and energy inputs from the moment the feedstock is collected to the final use or sale of the biochar. A robust LCA for a biochar project should cover:

- A baseline scenario against which to compare the emissions
- Feedstock sourcing: Emissions from collection, processing, and transportation of biomass.
- Processing phase: Energy consumption during pyrolysis/gasification, as well as emissions from equipment, combustion, and handling.
- Biochar physico-chemical analysis
- By-products: Accounting for co-benefits (e.g., renewable heat or bio-oil) and any associated emissions reductions.
- End use of biochar: Whether it's applied to soil, used in construction, or stored, and how long the carbon is expected to remain stable.
- Avoided emissions: GHG emissions that would otherwise be emitted if the biochar was not produced (e.g., if biomass would have otherwise decomposed anaerobically in landfills or lagoons).

Why is LCA important in reporting?

- It quantifies net climate benefit, supporting the issuance of carbon credits.
- It demonstrates transparency and scientific rigor to stakeholders and buyers.
- It helps identify opportunities to optimize processes and reduce emissions further.

Many carbon credit methodologies now require LCA-based verification, including third-party validation and periodic updates.

Verification

A qualified third-party verifier examines the project's documentation, site, and laboratory results. This independent review confirms whether the project meets the criteria of the chosen certification standard and if the data are real and if the number of carbon credits claimed is true. Otherwise, either the project is rejected, or the number of carbon credits are adjusted. Successful verification leads to the issuance of carbon credits.

Registry and Transparency

After verification, carbon credits are registered in an official registry (such as Puro.Earth, Verra, or EBC). This makes the credit traceable and avoids double-counting. Registries also ensure that each credit is only used once. The buyers feel more secure about the real carbon sequestration they purchase so that they can officially claim they reduce their emissions.

Table 1. Example of questions to be asked for biochar carbon credits project development

Step	Questions
Eligibility and registration	What are the eligibility requirements? How can I become eligible? Do I need to modify my production to be eligible?
Project development	What documents do I need to provide? How do the project documentation be filled? What data have to be collected? What supporting documents are needed?
Verification and certification	How do I mitigate risks? How do can I prepare for auditing? What are the auditors look for? What documentation need to be provided? What is the process and the timeline to obtain final certification?
Sales and marketing	What is the marketing strategy? Are the co-benefits well identified and explained? How much time and effort does it require? Who in my team and my network has the time and the specialization to do so? Who and how to sell to? What is good pricing, payment and contract terms?
Revenues	Do I obtain the max cash for my credits? What type of account do I need? What are the financial risks? Can I pre-sell my credits? What type of long-term agreement can I get?

Effective MRV systems should be designed early in the project and refined over time. Digital tools, remote sensors, and regular updates can help maintain high-quality data and streamline future verifications. Table 1 summarizes the key questions that project developers should address throughout the development of a biochar carbon credit project, from eligibility and registration to verification, certification, and commercialization.

4. Investment, Revenues, and Risks

Capital and Operational Costs (CAPEX and OPEX)

Setting up a biochar carbon project requires upfront investment. This includes equipment (e.g., pyrolysis units), land, buildings, and installation. The CAPEX varies from a few hundred dollars to 100 million dollars. The low cost is often associated to Kon-Tiki kiln. The most expensive ones are usually installed in developed countries for large quantities of feedstock (200000 t/yr) fully automated with a high control on the production and the quality of the products. Batch equipment is cheaper but require more labor than fully automated.

Operational costs (OPEX) include labor, feedstock transport, maintenance, energy, certification fees, laboratory analyses, and monitoring. Projects in remote regions may face higher costs due to infrastructure or logistics limitations.

Revenue Streams

You can generate revenue through:

Biochar sales: More than 60 markets exist for biochar, but only a few can access carbon credits. To access carbon credits, the biochar can't be burn, sold to retail markets, and a proof of its location and usage must be provided. As for example if use in litter, then it must be proven that the litter go to the soil at the end of its use. Soil amendments, animal feed, concrete, cement, asphalt and construction materials are eligible for carbon credits. It is frequent that a production facility sells part of its biochar to application eligible for carbon credits and other part for other types of usages.

Several plants generating carbon credits produce biochar as a co-product. This could be a biomass power plant, a bio-oil production facility with biochar being a waste.

Carbon credits: Verified credits can be sold for prices ranging from \$50 to over \$300 USD/tCO_2e depending on the buyer, quality, and certification standard and also on the duration of its potential life (GECA & Accend 2022; Puro.Earth Market Reports 2024). We often see carbon credits range of 150-200 USD/tCO_2e.

Co-products: Energy (syngas, heat, electricity), wood vinegar, bio-oil, and CO_2 in some systems could also be sold.

The return on investment (ROI) depends on the size and the type of project. Small projects may break even in 2–4 years, while larger industrial-scale projects may take longer than 7 years, but offer higher long-term profits. Carbon credit revenue is frequently the most significant financial driver or equal to the sell of the biochar or the bio-oil.

Financial Risks and Mitigation

Launching and operating a biochar carbon credit project involves significant financial uncertainty. Understanding the main financial risks and preparing mitigation strategies is essential for long-term viability.

1. Volatile Carbon Markets

Risk: Carbon credit prices in voluntary markets can fluctuate significantly depending on global demand, investor confidence, and shifts in corporate climate strategies. A price crash or reduced demand for biochar credits could undermine your expected revenue streams.

Mitigation Strategies: Diversify revenue streams: Don't rely solely on carbon credits—ensure that biochar, heat, or by-products (like bio-oil) also contribute to income.

2. Additionality

Additionality is a central requirement in carbon markets and a common stumbling block that can jeopardize the eligibility of a biochar (or any carbon) project for issuing carbon credits.

- **What is Additionality?:** In the context of carbon credits, additionality means that the carbon removal or emissions reduction would not have happened without the carbon credit incentive. In simpler terms: the project must go beyond business as usual and only exist or succeed because of the revenue or support from carbon credits
- **Why is Additionality Important?:** Carbon credits are supposed to fund new climate action. If a project would have happened anyway (e.g., because it's already profitable or legally required), issuing credits would not represent a *real* climate benefit—it would be double-counting or greenwashing.
- **Registries like Verra, Puro.Earth, and others require projects to prove that:**
    - The project isn't legally required.
    - The project isn't already financially viable without credit revenue.
    - The emissions/removals are new and real.
- **How Additionality can Become a Risk?:** If a project fails to meet additionality criteria, it can:
    - Be rejected during validation or not registered.

    - Lose eligibility partway through the crediting period.
    - Be challenged by auditors, NGOs, or the registry—reducing credibility or causing credit reversals
- **This is especially risky when:**
    - The project receives public subsidies that make it financially viable without carbon revenue.
    - The project uses a technology that's already common in the region (seen as business-as-usual).
    - The biochar is produced from feedstock that has existing high-value uses (e.g., already used as compost or fuel).
    - The carbon credit income is a minor portion of the total revenue, making it hard to prove it was a key incentive.
- **Mitigation strategies: Protect your project's eligibility:**
    - Clearly demonstrate in your documentation that carbon credit revenue is essential for the project's feasibility.
    - Show that no law or regulation mandates the activity.
    - Use marginal, low-value, or waste feedstocks (e.g., forest residue, crop waste) that wouldn't otherwise be processed.
    - Include a financial additionality analysis (costs vs. projected revenue) when submitting to registries.
    - Engage with a consultant or registry early to pre-assess additionality risk.
- **Example of high additionality:** A small-scale biochar project using forest waste in a remote region, where no other infrastructure exists, and carbon credit income is the main funding source.
- **Example of low additionality:** A large sawmill in Europe starts producing biochar from its waste but is already profitable and receives state subsidies. If carbon credits are just an extra revenue stream, the project might be ineligible.

A difficulty in diversifying the revenue concerns the additionality.

- **Long-term offtake agreements:** Secure forward contracts with buyers (e.g., corporate buyers or climate funds) to lock in credit prices.
- **Sell in multiple markets:** Work with registries that allow sales on several platforms or connect to large corporate buyers (e.g., Puro.Earth, Carbonfuture, Patch).
- **Build financial buffers:** Include conservative assumptions in your business model and maintain reserves to absorb market dips.

3. Certification Delays

Risk: Verification, auditing, and credit issuance processes—especially with first-time projects—can take longer than expected. Delays may result from incomplete documentation, evolving methodologies, or backlogs at registries.

Mitigation Strategies: Choose experienced partners: Work with project developers and consultants familiar registry. They will help identifying the right registry for your project and fill up all the requires.

Plan for long lead times: Factor in 6–12 months or more from project start to first credit issuance in your financial model. For some registries, it may take up to 2 years.

Invest in strong MRV systems: Good monitoring, reporting, and verification processes reduce the risk of rejections or review delays. Not planning the MRV, data acquisition and proof early in the project is a frequent mistake delaying the process and increasing the overall expenses related to building a carbon credit project.

Start small and scale: Consider piloting with a small facility or limited volume to smooth out the learning curve before going full scale.

4. Policy Shifts

Risk: National and international carbon regulations, subsidy programs, or carbon accounting rules change very fast over time. These changes can affect project eligibility, financial incentives, or reporting obligations – sometimes retroactively. Make sure to select a consultant that is closed to your registry so that he can help you adjust quickly and efficiently.

Mitigation Strategies: Monitor policy landscapes: Stay updated on local and international developments in climate and carbon markets. Join industry groups like IBI (International Biochar Initiative) or local biochar or clean tech clusters.

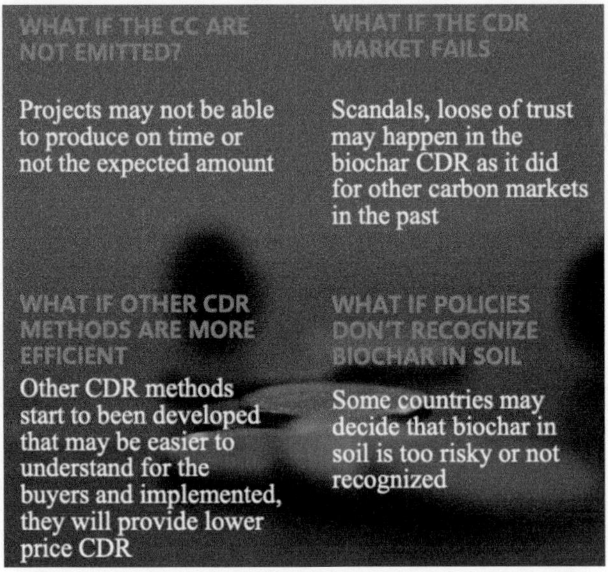

Figure 3. Is biochar really a solution for carbon sequestration? What are some of the risks?

Design flexible systems: Ensure your operations can adapt to new requirements—e.g., co-registering credits, changing feedstocks, or adding additional carbon sinks.
Build conservative forecasts: Base revenue models on the lowest foreseeable price and limit dependence on uncertain future policies.
Diversify geographically: If possible, work in regions with different regulatory environments to spread policy risk. Figure 3 illustrates the main risks associated with biochar carbon credit projects and highlights why risk identification and mitigation are critical to the long-term viability of these projects.

5. Timeline and Expectations for Project Development

Table 2 presents the typical development stages and indicative timelines for biochar projects, noting that several activities may occur concurrently depending on project scale and context.

While Table 2 focuses on the physical development of biochar production projects, Table 3 outlines the specific timeline associated with the development, auditing, certification, and monetization of biochar carbon credits project development.

Table 2. Development timeline stages for biochar projects (Most of them can be concomitant)

Step	Description	Timeline
Feedstock sourcing	Identify, test, and secure suppliers	1–6 months
Technology selection and procurement	Evaluate systems, order, and receive equipment	3–12 months
Market development for biochar	Identify, evaluate, obtain letters of interest	6–18 months
Financing	Identify financial sources and sign	6–12 months
Facility setup and operations	Obtain permits, build infrastructure and conduct start-up tests	12–36 months

Table 3. Development timeline stages for biochar carbon projects

Step	Duration (months)	Comment
Preparation	2 months	This is the preparation section, setting up data collection methods and operations.
Physical project build	2–5 years	The carbon project requires the physical project to be built and operational. This timeline is determined by the Producer of biochar (see previous table).

Carbon project development	3–6 months Based on the standard selected + data collection capacities	Generally, the carbon project development is started during the physical project build for best, accurate results, but can also start before with engineering. It can however only be completed once commissioning is over, because it depends on operations for data.
Commissioning of plant	TBD Generally, 1–3 months when it goes well	Timeline determined by the Client. Delays in this will result in delays for the overall carbon project.
Auditing	1–6 months Depends on the standard	Depends on auditor timelines, availability, standard and build of the auditing package, precision and accuracy of data. Certain standards are becoming slow in setting time for visiting the plant for the auditing.
Certification	0–3 months after auditing	Following auditing results, standard may request reviews of the project internally or additional auditing, but generally if the audit is passed then certification is issued shortly after.
Carbon credit monetization	Ongoing over several years Can start during construction or before, but bulk of the work is after auditing	Major activity will depend on the strategy determined for commercialization.

6. Challenges, Stakeholder Engagement, and Practical Solutions

Challenges

Regulatory challenges: Different countries have different rules on waste handling, emissions, and carbon reporting. Permitting for pyrolysis units may be unclear or slow.
Technical challenges: Inconsistent biochar quality can affect certification. Unoptimized processes can increase emissions or reduce carbon yields.
Market access issues: Selling biochar or credits without market connections is difficult. Many buyers want verified, premium-grade credits.

Solutions

Figure 4 provides an overview of key mitigation pathways and stakeholder-driven solutions to address the principal challenges faced by biochar carbon credit project development.

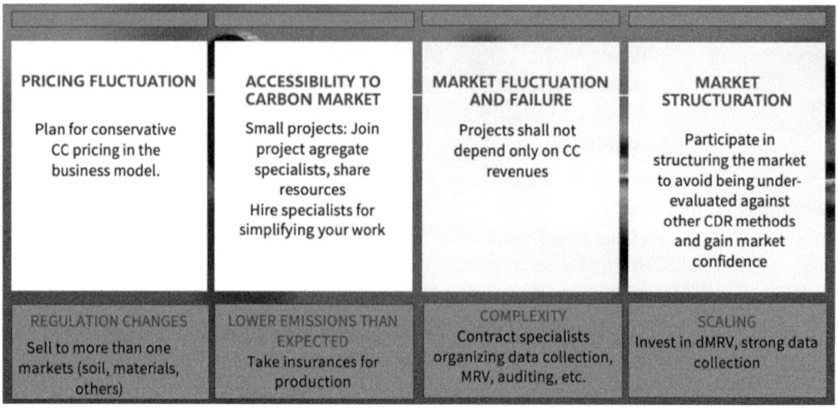

Figure 4. How to mitigate challenges?

- Use certified, well-tested technologies.
- Form partnerships with local industries or NGOs.
- Join carbon project networks and attend global events.
- Stakeholder engagement: Engaging local farmers, municipalities, and policymakers increases social acceptance and can provide stable supply and demand chains. Projects that include community participation tend to last longer and perform better.

7. Case Studies: Lessons from Successful Biochar Carbon Projects

Case 1: Large Power Plant in the US

- Feedstock: Wasted wood from the forest
- Tech: Biochar as a byproduct from energy production with relatively low capital cost compared to the power plant
- Timeline: 18 months using the services of a specialist in biochar
- Credits sold to large corporations via a broker to voluntary buyers
- Impact: Maintained local jobs and the biomass power plant without which it would close
- Challenges: First of its kind to be certified – delayed because of a lack of understanding by the standard, solved with the knowledge of the specialist. Explosion of a reactor – delaying carbon credits delivery, causing problems with buyers of pre-sold carbon credits – resolved with the network of the service provider by selling carbon credits from other projects of the specialist.

Case 2: Small-scale Biochar Project on a Small Island

- Feedstock: Invasive plant and storm clean-up
- Tech: Small semi-mobile continuous pyrolysis, relatively low capital costs

- Timeline: 24 months
- Credits pre-sold to large corporations via a broker to voluntary buyers
- Impact: Create new jobs, clean the environment
- Challenges: Hurricane destroyed all on the island. Project highly delayed along with refinancing difficulties. The selected standard changed its rules by not accepting small projects anymore. The producer potentially not able to quality for this new rule, may have to work with another standard – meaning doing again and adjusting the LCA, PDD, MRV and putting more money into the project development.

Case 3: Agricultural Biochar Project in the US

- Feedstock: Farm wastes
- Tech: Fully automated medium size system
- Timeline: 18 months using the services of a specialist in biochar
- Credits sold to small and large corporations via a broker to voluntary buyers
- Impact: Solve a problem with local farm wastes, generate local jobs
- Challenges: First of its kind to be certified – delayed because of a lack of understanding by the standard. Lack of understanding the importance of high-quality data. All solved with the knowledge of the specialist. The standard did not have time for auditing – delayed by 5 months. Resolved with the strong relationship between the specialist and the standard. The project generates only a few carbon credits with low financial return, resolved by multiplying the production plants and the co-benefits.

Key Lessons Learned

- Be patient, insistent, and realistic with the timeline: 18–60 months for overall project development from ideation to production of biochar and carbon credits first sales, which includes 12–24 months for carbon credits project development
- Expect the unexpected
- Prepare early for certification documentation
- Make it easy for the buyers of your carbon credits, develop the best data collection and prove them, be consistent in your production and your data set
- Build strong local partnerships along with the support of a specialist in biochar project development and another one in biochar carbon credits
- Multiply co-benefits for increasing carbon credit value and multiply products
- Don't sell carbon credits before emission unless you're 99% sure to produce them.

8. Future Trends in Biochar Carbon Markets

As the climate crisis deepens and global pressure to cut emissions grows, the biochar sector is evolving rapidly. Several key trends are shaping its future and increasing its credibility and accessibility in carbon markets.

Policy Inclusion: Governments are starting to recognize biochar as a valid method for removing CO_2 from the atmosphere. Currently, most biochar carbon credits are traded on voluntary markets. However, major regulatory systems such as the European Union Emissions Trading Scheme (EU ETS) are now reviewing how biochar could be integrated into their compliance markets. If accepted, this would allow companies legally required to reduce emissions to buy biochar credits, greatly increasing demand. Other governments use financial incentives (US government with its tax reduction of the 336 program and other programs for improving carbon in soils). We expect the biochar carbon market to double every 5–7 years. However, some countries like the Canadian government may decide that retract from the carbon taxes, reducing the needs and trend of the carbon market. Figure 5 provides empirical context for the discussion on future trends by showing the increase in certified CDR projects over time, reflecting growing market confidence and institutional recognition.

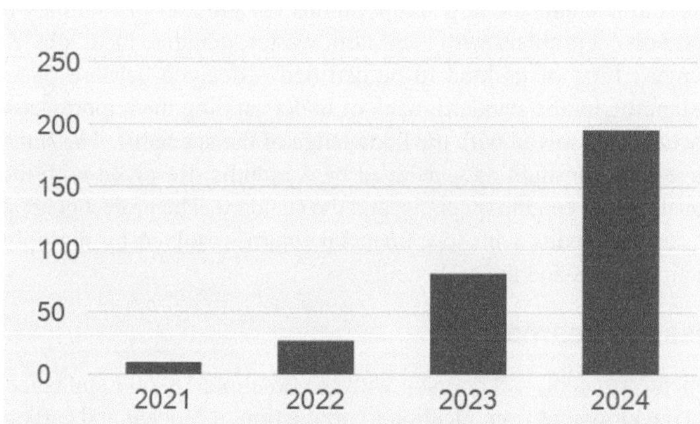

Figure 5. Number of certified CDR projects by Puro.Earth over time (data from Puro.Earth website).

Tech Improvement: Technology providers take experience over time. They become more reliable. Emission controls are improving, which helps ensure biochar production has a low environmental impact. Innovations also focus on energy recovery – such as capturing syngas to heat the reactor – which improves both climate performance and economic viability. New pyrolysis technologies, modular or mobile, become available that are smaller, cheaper, and cleaner. These systems can be installed in remote areas or small farms, allowing decentralized biochar production.

Financiers: They are gaining a better understanding of biochar and its production systems, and are becoming more confident in the carbon credit market. As trust grows, it becomes easier for project developers to access funding – a trend expected to strengthen over time.

MRV Automation: One of the biggest challenges in carbon credit projects is MRV. More biochar projects will use digital sensors, remote monitoring, and artificial intelligence (AI) to automatically track production, energy use, emissions, and carbon content. These tools reduce human error, speed up verification, making it easier for all size of projects to enter the market. But MRV automation requires investment.

Blockchain and Traceability: Buyers of carbon credits want to make sure that their credits are real, unique, and not used twice. Blockchain technology provides a secure and transparent way to track carbon credits from the source (biochar production) to the buyer. Some platforms are already using blockchain to store MRV data and issue "tokenized" carbon credits that are easier to trade online. The platforms for managing carbon credits and their dataset are improving each day.

Soil Regeneration: Farmers are beginning to see biochar not only as a climate tool but also as a soil amendment that enhances long-term productivity. This dual benefit makes biochar more attractive to agricultural investors and could open new funding and subsidy opportunities for farmers. However, very most of the agricultural actors in the world have not yet heard about biochar and even less about the carbon credits related to it. Education is needed but the trend shows a large increase in this awareness throughout the planet.

9. Conclusion

Biochar carbon credit projects offer an exciting opportunity to turn waste into value. With the right approach, these projects can be profitable, climate-positive, and socially impactful.

Recommendations

1. Build up a team with knowledge on project development, financing, technological aspects
2. Partner with other entities, together covering the feedstock provider, the potential user of the biochar and either the technology provider or the financing or the operating team.
3. Start by learning about different aspects of biochar projects and use a consultant specialized in these types of projects, who can help with feedstock, tech, and market selection, business plan and financial model.
4. Start with a pilot to test and learn on feedstock, tech, and markets.
5. Plan the unexpected and be realistic about timing. It takes longer than you think.
6. Join standard bodies early and study their methodologies.
7. Use third-party consultants for the carbon credit project development, MRV, certification, and the sales of carbon credits
8. Stay updated on policy changes and innovation.

References

Carbon Offset Guide. https://www.offsetguide.org

Carbon Standards International (CSI). https://carbonstandards.org

Climate Action Reserve (CAR). Biochar Protocol Development. https://www.climateactionreserve.org

European Biochar Certificate (EBC). https://www.european-biochar.org

European Biochar Industry Consortium (EBI). https://www.biochar-industry.com

GECA Environment and Accend. The Biochar Carbon Market: Two-years-old and Booming. 2022. https://www.gecaenviro.com

International Biochar Initiative (IBI). https://biochar-international.org

Ithaka Institute. https://www.ithaka-institute.org

Label Bas Carbone France (LBC). https://www.ecologie.gouv.fr/label-bas-carbone

Puro.Earth. CORC Standard and Registry. https://puro.earth

US Biochar Initiative (USBI). https://biochar-us.org/

Verra. Biochar Methodology Draft. 2021. https://verra.org

Woolf, D., Lehmann, J., Ogle, S., Kishimoto-Mo, A.W., McConkey, B., and Baldock, J. 2021. Greenhouse gas inventory model for biochar additions to soil. *Environmental Science & Technology* 55(21): 14795–14805. https://doi.org/10.1021/acs.est.1c02425

Section 3 – Shaping Tomorrow with Biochar: Policies, Education, and Real-World Insights

Biochar knowledge is reaching the most remote corners of the world, thanks to the dedication of NGOs and countless champions spreading insights into multiple languages. This knowledge is vital for advancing biochar policies and fostering widespread adoption, especially in developing countries where such policies are still emerging. While challenges remain, high production costs, unclear standards, limited education, and carbon credit pricing have limited biochar's potential to transform economies, enhance sustainability, and create innovative business opportunities. Despite these challenges, successful companies around the world are already using biochar in diverse applications and leveraging carbon credits, demonstrating that thriving biochar businesses are possible even in the face of barriers.

KEY LESSONS

- Education is the spark that ignites biochar's promise—sharing knowledge empowers communities, transforms practices, and multiplies impact.
- A bold, strategic policy approach can unlock biochar's full potential, making meaningful contributions to sustainable development.
- Biochar holds transformative power across industries, but it requires accessible production, clear standards, informed communities, and fair carbon credit pricing.

CHAPTER

7

Biochar Public Policy in Developing Countries and Its Roles in Innovation

Shanthi Prabha Viswanathan[1*] and Bluvin Ravindran[2]

[1] Biochar Today, Felica Mosca, Florida, US

[2] Super Biochar, Norwalk, USA

1. Introduction

At the nexus of climate change, mitigation, and sustainable development, biochar represents a unique and powerful tool with particular relevance for developing countries. Produced through the pyrolysis of biomass material, biochar offers a highly stable carbonaceous matrix and a proven, versatile benefit path to address global challenges. Studies, experiments, and real-world applications showcased the benefits of biochar, including improved soil fertility and yield optimization, sustainable waste management, and the appropriate and approved technology for maximum soil carbon sequestration. As approved by the Intergovernmental Panel on Climate Change (IPCC), (*U.S. Biochar Coalition Response to UNFCCC Article 6 Call for Comments* 2023) biochar is widely recognized as a viable and relatively affordable Carbon Dioxide Removal (CDR) solution accounting for an estimated 94% of delivered carbon removal credits in 2023.

Despite this immense potential, systemic barriers hinder the scalable adoption of biochar. A primary challenge is the absence of a harmonized and robust policy network, as regulatory ambiguities, such as considering high-grade biochar as waste, discourage investment and limit market development. Financial obstacles, including the high upfront cost of production and the complex, unaffordable nature of carbon credit certification, excessively affect smallholder farmers. Additionally, a general lack of awareness and consistent knowledge impedes global acceptance of biochar. To overcome these obstacles, a strategic,

*Corresponding author: shanthiprabhav@gmail.com

DOI: 10.1201/9781003467946-10

multifaceted approach is inevitable. A firm public policy should act as a catalyst, offering regulatory clarity and financial assistance to de-risk investment and promote innovation. The central argument of this chapter is that public policy is the essential catalyst needed to bridge this gap, transforming biochar from a niche product into a scalable solution that can drive economic development and bolster climate resilience in the Global South. A critical policy vacuum exists in many developing nations, where the biochar industry struggles with a lack of demand, fragmented supply chains, and limited access to financing.

This chapter provides a thorough overview of the current policy environment for biochar, drawing from research and regional examples in developing countries. It also examines the increasing demand for biochar credits within compliance carbon markets, such as the European Union Emissions Trading Scheme (EU ETS) and new initiatives in South Africa, to extract key lessons from developed nations' policy and market advancements. The ultimate aim is to create a practical framework that helps policymakers bridge the divide between biochar's scientific potential and its real-world application.

2. Environmental and Economic Significance of Biochar for Developing Countries

Biochar offers a powerful economic and environmental tool for developing nations by boosting agriculture and creating new revenue streams. Globally, most developing economies are characterized by extensive agriculture sectors and related production or generation of agro-waste biomasses, enabling them to leverage biochar technology for sustainable development (Kannan et al. 2013). Developing countries offer substantial biochar production potential from agricultural waste. Countries like India generates 500–550 Mt of crop residues annually, with 98.4–131.9 Mt currently burned, contributing to GHG emissions (Bhuvaneshwari et al. 2019). Utilizing this waste for biochar can combat climate change, enhance crop yields, and address deforestation and rural energy needs. Similarly, East Africa's 33.3 Mt of maize residues and Ghana's diverse biomass resources present significant opportunities for sustainable biochar implementation (Siddiqui 2025).

Biochar offers a powerful economic tool for developing nations by boosting agriculture and creating new revenue streams. By improving soil health, biochar can increase crop while cutting the need for expensive synthetic fertilizers, which enhances food security for smallholder farmers (Nassary 2025; Afshar and Mofatteh 2024a). Production turns agricultural waste into a valuable product, generating direct income and reducing harmful open-air burning (Yu et al. 2024). The sale of carbon credits from biochar can further increase farmer income by as much as 120%, often becoming the primary source of revenue for small-scale projects (*Sustainability of Biochar Systems in Developing Countries – International Biochar Initiative* n.d.; *How Countries Can Harness Biochar in the*

Fight Against Climate Change n.d.). With versatile uses beyond agriculture, such as in livestock feed and construction, and the ability to generate clean thermal energy, biochar presents a flexible and decentralized economic model for rural communities (Kumar et al. 2020).

Closed-loop biochar production systems present enhanced opportunities. Specifically, decentralized biochar production approaches can promote local job creation, improve waste handling, and increase resource efficiency within a circular bioeconomy (Yaashikaa et al. 2020). Moreover, biochar supports several Sustainable Development Goals (SDGs), such as No Poverty (SDG 1), Zero Hunger (SDG 2), Affordable and Clean Energy (SDG 7), Climate Action (SDG 13), and Life on Land (SDG 15), illustrating its wide-ranging contribution to comprehensive development (Babu et al. 2023).

Research consistently demonstrates biochar's integrated benefits across agriculture, waste management, and climate change mitigation, as depicted in Fig. 1. These are interconnected systemic challenges in developing countries. For example, burning agricultural waste degrades soil, pollutes the air, and worsens climate change, exacerbating agricultural vulnerabilities. Biochar acts as a synergistic intervention, transforming waste into a carbon-sequestering soil amendment, offering a more impactful solution than isolated efforts.

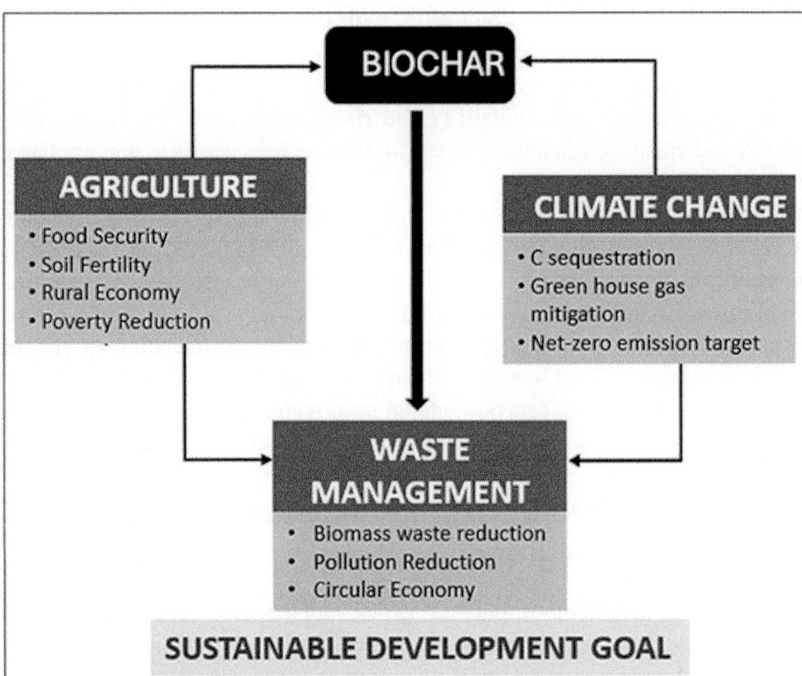

Figure 1. Relevance of biochar for developing countries.

Therefore, biochar presents a viable solution for fostering sustainable agricultural practices, waste management, and environmental stewardship (Senadheera et al. 2025). Moreover, the integration of biochar into waste management and bioenergy systems aligns it with the principles of a circular bioeconomy (Waheed et al. 2025). Biochar's versatility aligns with sustainable development, offering a valuable solution for climate change, soil degradation, and pollution. Pyrolyzing residues to produce biochar can prevent 66–130 billion metric tons of CO_2 equivalent emissions over a century, with 50% from carbon sequestration, 30% from fossil fuel substitution by pyrolysis energy, and 20% from reduced methane and nitrous oxide emissions (Vuppaladadiyam et al. 2023). The biochar industry, valued at $541.8 million in 2023, is set for significant growth, projected at a 13.9% annual rate from 2024 to 2030 (Afshar and Mofatteh 2024a). Therefore, maximizing biochar's benefits for sustainable development requires a holistic, cross-ministerial policy approach, rather than confinement to a single sector.

Despite its potential, biochar's commercialization in developing nations faces significant financial and logistical challenges. There's a fundamental trade-off: simple, low-cost kilns are often profitable for individual producers, but more environmentally beneficial, advanced systems are typically too expensive to be economically viable (Owsianiak et al. 2021). This creates a paradox where the most profitable methods don't align with the most environmentally sound ones. The core issue isn't technological, but a market and policy failure. Many producers rely heavily on a single revenue stream from carbon credits, a model that is susceptible to price volatility and rising labor costs. Furthermore, the complex and costly certification process for these credits serves as a major barrier for many projects.

To correct these market failures, public policy should create a diversified market that rewards biochar projects for their full benefits, not just carbon sequestration. A single, one-size-fits-all approach is not effective. Policymakers should instead implement a two-tiered model. One tier would support low-cost, artisanal production with financial strategies like micro-financing to enhance local livelihoods and food security immediately. The other tier would incentivize larger, industrial-scale projects through public-private partnerships to meet global standards and achieve significant climate impact (Kochanek et al. 2022).

3. Navigating the Existing Policy Landscape: A Global and Regional Analysis

The regulatory treatment of biochar varies significantly across jurisdictions, creating a complex landscape for its widespread adoption (Pourhashem et al. 2019). Jurisdictions generally handle it in one of three ways: as an agricultural input, as waste, or with an ambiguous legal status. Some regions, such as Switzerland, have established formal regulatory pathways that recognize biochar's value as a soil amendment, making certification like the European Biochar Certificate (EBC) mandatory for agricultural use (Schmidt 2013). In contrast, a significant hurdle is its frequent classification under existing waste laws. If biochar is produced from materials considered 'discarded', it can be categorized as waste, which imposes costly and burdensome obligations on producers and ties it to waste management regimes instead of frameworks designed for resource utilization. This ambiguity between a valuable product and a waste material creates a significant policy challenge (Štrubelj and Singh Ghaleigh 2025). This regulatory inertia, which applies old frameworks to a novel solution, discourages investment and innovation (Table 1). A clear,

Table 1. Biochar classification and regulatory status in select countries/regions

Country/ Region	**Primary classification (general)**	**Specific regulations/Acts (examples)**	**Key implications of classification**
Switzerland (EU Aligned)	Agricultural Input/Product	European Biochar Certificate (EBC) (Obligatory for agricultural use)	**Enabling:** Clear product status, fosters market confidence, ensures quality and safety.
United Kingdom	Waste (if from discarded materials or co-product of energy)	Waste Framework Directive (retained EU law) (Štrubelj and Singh Ghaleigh 2025)	**Hindering:** Imposes costly waste management obligations, creates regulatory uncertainty, and disincentivizes scaling.
India	Ambiguous (potential as soil amendment, waste-to-value, climate tech)	Carbon Credit Trading Scheme (CCTS) (includes biochar as innovative tech in voluntary offset) (*India Regulatory Updates – 2024 Environment and Climate Change – Lawwiser* n.d.), Biomass Programme (Biomass Programme, Ministry of New and Renewable Energy n.d.)	**Evolving/Ambiguous:** Emerging recognition in climate/energy policies, but lack of explicit, unified legal definition for agricultural input status creates uncertainty.
China	Resource/ Agricultural Input	Ministry of Agriculture support for carbon sequestration and manure management (Mal et al. 2024)	**Enabling:** Government support for development and private sector participation, fostering the value chain.

enabling legal definition that recognizes biochar as a product is essential for attracting the necessary investment, scaling up operations, and delivering its full environmental and economic benefits (Anokye 2024).

The policy landscape for biochar in the Global South is a dynamic and evolving one (Jaschke and Biermann 2022), with various regions adopting distinct approaches based on their unique economic, social, and environmental contexts.

Africa: Biochar initiatives in Africa are often spearheaded by non-governmental organizations (NGOs) and international partnerships, focusing on grassroots, community-level implementation. Organizations like Solidaridad have launched programs in Ghana, Zambia, and Tanzania, where they train smallholder farmers to produce biochar from agricultural waste. This strategy not only improves soil health and crop yields but also generates a new source of income for farmers by selling biochar in the voluntary carbon market. This approach leverages readily available agricultural residues and directly benefits local livelihoods (Scholz et al. 2014). In a significant policy change, South Africa will modify its carbon tax regime from 2026 to officially recognize carbon dioxide removals (CDR) like biochar, which is the first among developing nations. This move is significant given the country's high unemployment and reliance on coal, as it positions biochar as a tool for creating jobs and reducing emissions (Carbon Pulse 2025; Manuell 2025).

Latin America: The biochar industry in Latin America is growing quickly, projected to rise from US$ 197.5 million in 2023 to US$ 864.6 million by 2033 at a 15.9% CAGR. Supportive government policies on sustainable agriculture, carbon neutrality, and waste management encourage the use of biochar, alongside abundant regional resources such as crop residues and forestry by-products. Countries like Brazil and Mexico prioritize biochar in climate strategies, with initiatives linked to soil restoration, water conservation, and reduced fertilizer dependency. As policy frameworks increasingly align with climate and food security goals, biochar is set to become a key instrument in Latin America's sustainability agenda

In Latin America, biochar-related policies are still in a nascent stage, and there is no explicit mention of biochar as a soil enhancer in government plans. Instead, the industry receives indirect support through broader sustainable agriculture programs like Brazil's Low Carbon Agriculture Program (ABC Plan). The region has a high potential for biochar due to its abundant feedstocks, but it faces significant barriers. These include a lack of specific policies and programs, a considerable knowledge and empirical development gap, and a shortage of local laboratories for the required characterization tests for international certification. This leads to a persistent climate finance gap and limits the ability to effectively scale the technology (Carlino et al. 2022).

South East Asia: Regarding Southeast Asian countries, stand-alone biochar policies are lacking but are increasingly integrating biochar into broader climate change, crop burning reduction, and sustainable agriculture strategies. The ASEAN Guidelines on the Reduction of Crop Burning (adopted at the 46th AMAF in 2024) highlight converting agricultural residues into biochar as a sustainable alternative to open burning, linking it to soil fertility improvement, carbon sequestration, and greenhouse gas mitigation. Supported by ERIA, JIRCAS, NARO, GIZ, and CABI partners, ASEAN (links appended) explores case studies, technical catalogues, and inventory assessments for biochar initiatives across Member States. These steps position biochar as part of ASEAN's wider Food, Agriculture, and Forestry Sectoral Plan (2026–2030), with implementation expected to be advanced through crop insurance, financial schemes, and public-private partnerships (ASEAN Main Portal n.d.). Southeast Asian policymakers promote biochar adoption by aligning legislation, developing shared infrastructure, and coordinating carbon markets to enable cross-border credit exchange (Illuminem 2025). This regional approach leverages pooled resources and expertise, offering a more effective alternative to isolated national strategies.

India: Most developing countries' biochar landscapes are evolving, especially in India, with increasing recognition of its potential across various sectors. The government promotes sustainable agriculture and waste-to-energy initiatives through policy support, including carbon credit programs, subsidies, and grants. The Ministry of New and Renewable Energy's (MNRE) Biomass Programme indirectly supports biochar by offering financial assistance for biomass processing, which can facilitate its production as a co-product (Waste to Energy Programme | Ministry of New and Renewable Energy | India, 2020). The India BioChar and BioResources Network (IBBN) actively advocates for state policies and public funds to bridge the gap between scientific knowledge, policy, and implementation (*Policy – India BioChar and BioResources Network* n.d.).

High-level political endorsement is growing, exemplified by the Biochar Centre of Excellence inauguration at Kanha Shanti Vanam, Telangana, in India. In partnership with PayPal, this initiative aims to train rural entrepreneurs and decentralize biochar production (Business Minutes 2025). Furthermore, biochar can be strategically integrated into existing governmental schemes such as the Soil Health Card Scheme, the National Mission for Sustainable Agriculture (NMSA), and the National Biofuel Policy (NBP), enhancing soil health, promoting a circular economy, and ensuring its benefits are leveraged across multiple sectors. A pivotal development is introducing biochar as an innovative technology for greenhouse gas removal within the Carbon Credit Trading Scheme (CCTS), a national market-based mechanism. This move is significant as it transitions biochar from implicit support to explicit, economically driven policy. Successful projects, such as The Climate Foundation's biochar reactor for solid waste management in India, demonstrate its practical application. While India's biochar policy has been fragmented, the inclusion in the CCTS provides a clear financial incentive and regulatory clarity, which could unlock private investment and accelerate the scaling of biochar across the country (*Policy: Policies from Central and State Governments* 2018).

Apart from India, while federal biochar policies in other developing countries are not extensively documented, China stands out with its Ministry of Agriculture actively supporting biochar for carbon sequestration and manure management, and encouraging private sector involvement (Table 2). Chinese universities and

Table 2. A comparative analysis of regional biochar policy approaches

Country	Africa	Latin America	Southeast Asia	India
Feature	Grassroots and pockets of national policy	Indirect and nascent	Collaborative and harmonizing	Emerging compliance market
Primary Policy Approach	NGO-led, small-scale programs. Pockets of national-level initiatives (e.g., South Africa's carbon tax)	Indirect support through broader agricultural policies (e.g., Brazil's ABC Plan)	Regional collaboration and knowledge sharing. Focus on harmonizing policy and reducing crop burning	Incorporation into existing policies (e.g., organic farming). Emerging compliance and voluntary carbon markets
Key Drivers	Climate adaptation, food security, and rural livelihoods. Job creation (South Africa)	Sustainable agriculture, soil health, and utilizing abundant biomass residues	Reducing crop burning, agricultural sustainability, and meeting regional net-zero goals	Addressing land degradation, crop burning, and meeting net-zero goals
Primary Barriers	Lack of national-level data and funding support. Over-reliance on a single revenue stream from carbon credits	Lack of specific biochar policies. Nascent carbon markets. High cost and logistical barriers to certification	Lack of infrastructure and consistent financing for projects. Need for regional policy harmonization.	Lack of standardized production protocols. Low awareness among farmers and industries. Need for investment
Noteworthy Initiatives	Solidaridad programs in Zambia, Tanzania. South Africa's carbon tax modification	Brazil's ABC Plan. BNDES funding for reforestation projects	ASEAN-level events promoting biochar as an alternative to crop burning	Carbon Credit Trading Scheme (CCTS) with a voluntary offset mechanism. Green Credit Programme

companies also engage in "South-South cooperation" by collaborating with ASEAN countries and the Asian Development Bank, providing training and exploring regional waste streams for biochar production. This trend demonstrates that developing nations are not solely dependent on North-South technology transfer; they are creating their expertise and sharing solutions with peers, highlighting an opportunity for countries like India to form similar partnerships to accelerate global biochar adoption (Mal et al. 2024).

4. The Carbon Market Evolution from Voluntariness to Compliance

Biochar has become the dominant technology in the voluntary carbon market, delivering an estimated 94% of all carbon removal credits in 2023 due to its long-term carbon permanence and relatively low cost compared to other engineered removal methods (Chiaramonti et al. 2024). Its credibility rests on robust third-party certification methodologies, with registries such as Puro.earth and Verra setting standards for quantification and monitoring. Endorsements from the Integrity Council for the Voluntary Carbon Market (ICVCM) have further strengthened buyer and investor confidence, making biochar one of the most trusted options for durable carbon removal. A new frontier is emerging as biochar seeks entry into compliance markets. In Europe, the biochar industry is advocating for its inclusion in the EU Emissions Trading Scheme (EU ETS) during the 2026 review, though policymakers are weighing risks such as 'abatement deterrence' and unsustainable biomass use, suggesting a cautious integration with safeguards (Malins et al. 2023).

Meanwhile, South Africa is preparing to reform its carbon tax system in 2026 to recognize CO_2 removals and increase penalties for exceeding emission limits, offering a pioneering model distinct from cap-and-trade. If major compliance systems begin accepting biochar credits, it would create a government-backed global standard, unlocking large-scale investment opportunities. This would be especially significant for the Global South, where abundant agricultural residues and strong social and ecological co-benefits make biochar projects highly competitive and attractive for sustainable development financing (Salma et al. 2024).

5. Gaps and Barriers to Biochar Adoption

The widespread adoption of biochar in developing countries is impeded by a multifaceted array of legal, financial, knowledge, and social barriers. The fundamental issue is the lack of a clear definition, as biochar is often regulated under existing waste laws designed for disposal rather than resource utilization (Štrubelj and Singh Ghaleigh, 2025). This ambiguity and the absence of mandatory quality standards, despite international frameworks like the European Biochar Certificate (EBC) and and the International Biochar Initiative (IBI)

Biochar Standards (Arbestain et al. 2019) create a regulatory catch-22 that deters investment and consumer confidence. Financially, small-scale producers face a significant paradox: while biochar is a key player in the global carbon credit market, the high upfront costs for production units and the complex, expensive process of carbon credit certification make these incentives inaccessible primarily to the very farmers who are crucial for scaling the technology (Salma et al. 2024).

Overcoming the barriers to biochar adoption requires addressing significant knowledge gaps regarding its various applications and how it performs best under different conditions (Anokye 2024). It is crucial to increase research and development efforts and to ensure that all biochar production technologies and application methods are made openly accessible, which will encourage widespread use (Masud et al. 2023). A science-to-market disconnect exists, where inconsistent research findings and a lack of standardized, simplified monitoring, reporting, and verification (MRV) frameworks make it difficult for practitioners to trust and apply biochar effectively (Abu El Haija and Santos 2024). Crucially, social and institutional barriers amplify these challenges. Smallholder farmers often lack access to finance, and gender inequities mean that women, who are a significant part of the agricultural workforce, face disproportionate barriers to participation due to time poverty and limited access to resources (Bamanyaki 2020). Finally, institutional fragmentation, with different government ministries operating in soils, leads to disjointed policies and a lack of coordinated, comprehensive strategies. Addressing these systemic issues requires a holistic approach that includes creating clear legal definitions, designing inclusive financial mechanisms, simplifying MRV processes, and fostering a whole-of-government effort explicitly sensitive to social and gender inequities.

6. Climate Policy Integration

Despite biochar's recognized potential for climate change mitigation, its explicit integration into national climate policies is in its early stages. This potential-to-policy gap is highlighted by studies demonstrating its significant but largely untapped capacity. For example, a study in Norway estimated that converting local forestry waste to biochar could sequester an amount of CO_2 equivalent to 13–40% of the country's agricultural emissions (Tisserant et al. 2022). Similarly, with its even larger forestry resources in Finland, a study estimated that biochar produced from forestry waste alone could sequester 5 to 12 million tons of CO_2eq annually, sufficient to cover a large portion of the country's annual agricultural emissions (Jokubė et al. 2025). These examples underscore the substantial, yet often theoretical, potential of biochar that has yet to be formally integrated into climate policy documents, which tend to focus on broader mitigation categories (Jokubė et al. 2025).

While biochar is recognized for its ability to store carbon for centuries and reduce agricultural greenhouse gas emissions, there is a gap between this recognized potential and its formal inclusion in policy. For example a relevant study result illustrates the significant role biochar can play in world-wide CDR strategies, with carbon dioxide removal potential of 6.23 ± 0.24% of total GHG emissions in the 155 countries covered based on 2020 data over a 100-year timeframe, and more than 10% of national emissions in 28 countries. Concentrated regions of high biochar carbon dioxide removal potential relative to national emissions were identified in South America, northwestern Africa and eastern Europe (Lefebvre et al. 2023). However, the "potential-to-policy" gap is likely due to a lack of standardized data and methodologies for quantifying and reporting biochar's carbon removal contributions within national accounting frameworks. To bridge this divide, a concerted effort is needed to develop country-specific, high-resolution data on biochar's potential and to establish standardized, verifiable carbon accounting methodologies. This would give policymakers the confidence to formally integrate biochar into their climate commitments, attracting necessary investment and accelerating progress toward ambitious goals like India's 2070 net-zero target.

7. Role of International Donors and NGOs

International organizations and NGOs are crucial in advancing biomass volarization and biochar adoption in developing countries by filling critical policy and implementation gaps (Ouro-Salim et al. 2024). Organizations like the International Biochar Initiative (IBI) support research and projects by facilitating information exchange, developing proposals, and providing technical advice. They often focus on micro-scale systems, such as biochar cookstoves, to ensure that production directly benefits local populations and biodiversity. Additionally, philanthropic organizations and initiatives like BiocharLife empower smallholder farmers to produce biochar and generate carbon credits, providing direct financial benefits and reducing rural poverty (Sörman 2023). Through on-the-ground pilot projects, international donors and NGOs effectively act as "de-risking agents" for biochar technology. By demonstrating biochar's practical and economic viability, especially through carbon finance for smallholders, they create successful models that can inform and accelerate the development of broader national policies (Scholz et al. 2014). This grassroots approach helps bridge the "knowledge-policy-application gap", and governments should formalize partnerships with these actors to leverage their expertise for adequate and equitable scaling of biochar solutions.

8. Lessons Learned from Developed Countries

Developed nations provide a model for scaling the biochar industry by establishing a clear legal, financial, and market foundation (Salo et al. 2024; Prochnow et al. 2024).

In Europe, a legal and technical foundation is being established for the biochar industry through regulations such as the Waste Framework Directive (WFD) and the Registration, Evaluation, Authorisation and Restriction of Chemicals (REACH) regulation. The WFD clarifies the distinction between 'waste' and 'by-product', a crucial legal precedent that can reduce regulatory burdens and costs for producers (Pourhashem et al. 2019). Similarly, the formal inclusion of biochar in the EU's REACH charcoal dossier makes compliance a legal requirement for producers, importers, and users, serving as a model for ensuring product quality and safety. The International Organization for Standardization (ISO) and the German Institute for Standardization (DIN) are also working to standardize biochar, further solidifying the industry's technical foundation. While these standards are crucial for building consumer confidence, the challenge for developing countries is to adapt them, not simply adopt them, as the required cost and infrastructure can be prohibitive.

Governments in developed nations are moving to fill critical knowledge gaps and lower investment barriers. They support large-scale research through the US Biochar Policy Project and the proposed Biochar Research Network Act in the US, aiming to create a financially enabling environment. For example, the US Biochar Coalition advocates for biochar to be eligible for tax incentives, such as those included in the Farm Bill and various Investment and Production Tax Credits. This legislative and regulatory recognition de-risks private investment, which is essential for scaling the industry. The biochar industry is hindered by a chicken-and-egg dilemma, where low demand and uncertain profitability discourage investment, limiting the large-scale supply needed to meet potential demand. A key lesson is that public policy should not directly fund all projects but act as a risk mitigator and market enabler by creating a stable regulatory environment and connecting small producers to larger markets.

The global biochar market is growing, increasing industrial production and diverse applications. However, biochar-producing companies are predominantly located in economically developed countries. The Nordic biochar market, for example, is in its nascent stages, primarily driven by small companies that often consider biochar production a side business. Despite this, Europe's annual biochar production reached 75,000 tons in 2023, with Nordic countries contributing approximately 28% of the total. In addition to pyrolysis, the industry is also exploring gasification and hydrothermal carbonization (HTC) for production. China and the United States lead research and patent development, holding nearly 71% of total biochar-related intellectual. These trends show that while the market is still maturing, the foundations for a scalable, diversified, and innovative industry are being laid.

9. Recommendations and Pathways Forward

With a history spanning millennia, biochar used by the indigenous Amazonian people dating back over 2,000 years is a promise to enhance soil fertility. Studies and field trials robustly proved biochar application is an accepted strategy for an extended scale of carbon sequestration and removal due to its long-term stability in soils. Recent advancements pointed out the versatile properties, including its ability to retain nutrients and water, adsorptive removal of heavy metals and other toxic chemicals, microplastics, organic polutants etc., from various environmental matrices. Multifunctionality of biochar enables its application across multiple industries, from soil amendment and environmental remediation to metallurgical uses and as an additive in construction. The pyrolysis process also yields additional valuable products including energy-dense gases, bio-oils and distillates capable of generating heat and power. Transforming biochar into high-value carbon forms like activated carbon, carbon nanodots, nanotubes and sheets, or graphite for extended use in various sectors like energy storage and the automotive industry. Biomass-derived biochar thus presents a sustainable and renewable alternative to fossil-based carbons, which currently dominate these markets (Salo et al. 2024).

A multi-pronged approach is essential to fully unlock biochar's potential in developing countries, addressing policy, economic, research, and social dimensions. Policy frameworks must reclassify biochar as a product, not waste, and establish national quality standards aligned with international certificates (Štrubelj and Singh Ghaleigh 2025). Biochar should be explicitly integrated into national climate, agricultural, and waste policies. Economic mechanisms must be enhanced by simplifying carbon credit programs for small-scale producers and offering subsidies or tax benefits for production. Increased research, development, and capacity-building funding is needed to create robust and transparent monitoring, reporting, and verification (MRV) frameworks and scale up training programs. Finally, social inclusion and cross-sectoral coordination are prioritized, including designing gender-responsive programs and strengthening communication among government ministries. This holistic policy framework will address fragmentation, regulatory ambiguity, and financial and social barriers, enabling developing nations to achieve climate goals and support resilient agriculture.

10. Policy Recommendations

A strategic, multifaceted policy approach is essential to unlock the full potential of biochar and ensure its effective contribution to sustainable development. Key recommendations are as follows:

- **Integration into Climate and Environmental Policies:** Policymakers must formally recognize biochar as a tool for achieving climate targets, such as those within the EU's Land Use, Land Use Change and Forestry (LULUCF) regulations and the Paris Agreement. This includes establishing a definitive carbon dioxide price and shifting biochar from voluntary to compliance carbon markets.
- **Regulatory Clarity and Standards:** Establishing standards for biochar quality and end-uses is crucial to build consumer and investor confidence. This also

requires policies that enable the production and use of biochar co-products, such as pyrolysis liquids.

- **Financial Incentives:** Governments should support the biochar industry's early development through financial instruments like grants, subsidies, and investment-like feed-in tariffs. This helps overcome the chicken-and-egg dilemma where initial high costs and investment uncertainties hinder scaling.
- **Waste Management Policies:** Policies are needed to discourage the direct combustion of potential biochar feedstocks and to ban the landfilling of materials suitable for biochar production.

References

Abu El Haija, K., and Santos, R. M. 2024. Rethinking biochar's MRV systems: A perspective on incorporating agronomic and organic chemistry indicators. *Sustainable Chemistry* 5(4): 287–307. https://doi.org/10.3390/suschem5040020

Afshar, M., and Mofatteh, S. 2024a. Biochar for a sustainable future: Environmentally friendly production and diverse applications. *Results in Engineering* 23. https://doi.org/10.1016/j.rineng.2024.102433

Anokye, K. 2024. From waste to wealth: Exploring biochar's potential in energy generation and waste mitigation. *Cleaner and Circular Bioeconomy* 9(July), 100101. https://doi.org/10.1016/j.clcb.2024.100101

Babu, S., Singh, R., Kumar, S., Rathore, S.S., Yadav, D., Yadav, S.K. et al. 2023. Biochar implications in cleaner agricultural production and environmental sustainability. *Environmental Science: Advances* 2(8): 1042–1059. https://doi.org/10.1039/d2va00324d

Bamanyaki, P.A. 2020. Barriers and opportunities for gender-responsive climate-smart agriculture adoption in Northern Uganda. *In:* Climate Change, Agriculture and Food Security (Issue Fao 2014). https://cgspace.cgiar.org/handle/10568/109811

Bhuvaneshwari, S., Hettiarachchi, H., and Meegoda, J.N. 2019. Crop residue burning in India: Policy challenges and potential solutions. *International Journal of Environmental Research and Public Health* 16(5). https://doi.org/10.3390/ijerph16050832

Biomass Programme, Ministry of New and Renewable Energy. https://mnre.gov.in/en/bio-mass/

Carlino, M., Gogorza, A., and Torcuato Di Tella, F. 2022. Climate investment in the food and agriculture sector in Latin America. https://newclimate.org/resources/publications

Chiaramonti, D., Lehmann, J., Berruti, F., Giudicianni, P., Sanei, H., and Masek, O. 2024. Biochar is a long-lived form of carbon removal, making evidence-based CDR projects possible. *Biochar* 6(1). https://doi.org/10.1007/s42773-024-00366-7

Heartfulness Institute and PayPal unveil Centre of Excellence to Empower Rural Entrepreneurs. (n.d.)

How Countries Can Harness Biochar in the Fight Against Climate Change. (n.d.)

India Regulatory Updates – 2024 Environment and Climate Change – lawwiser. (n.d.)

Jaschke, G., and Biermann, F. 2022. The policy discourse on negative emissions, land-based technologies, and the Global South. *Global Environmental Change* 75. https://doi.org/10.1016/j.gloenvcha.2022.102550

Jokubė, M., Hyyrynen, M., Pihlainen, S., and Hyytiäinen, K. 2025. Economic feasibility of biochar for carbon stock enhancement in Finnish agricultural soils. *Carbon Management* 16(1). https://doi.org/10.1080/17583004.2025.2465328

Kannan, P., Arunachalam, P., Prabukumar, G., and Govindaraj, M. 2013. Biochar an alternate option for crop residues and solid waste disposal and climate change mitigation. *Academicjournals.Org* 8(21): 2403–2412. https://doi.org/10.5897/AJAR12.2083

Kochanek, J., Soo, R.M., Martinez, C., Dakuidreketi, A., and Mudge, A.M. 2022. Biochar for intensification of plant-related industries to meet productivity, sustainability and economic goals: A review. *Resources, Conservation and Recycling* 179(September 2021): 106109. https://doi.org/10.1016/j.resconrec.2021.106109

Kumar, M., Xiong, X., Wan, Z., Sun, Y., Tsang, D.C.W., Gupta, J. et al. 2020. Ball milling as a mechanochemical technology for fabrication of novel biochar nanomaterials. *Bioresource Technology* 312(June): 123613. https://doi.org/10.1016/j.biortech.2020.123613

Lefebvre, D., Fawzy, S., Aquije, C.A., Osman, A.I., Draper, K.T., and Trabold, T.A. 2023. Biomass residue to carbon dioxide removal: Quantifying the global impact of biochar. *Biochar* 5(1). https://doi.org/10.1007/s42773-023-00258-2

Mal, M., Zhu, H., and Johnson, F.X. 2024. Prospects and challenges for land-based climate change mitigation in support of carbon dioxide removal in China (Issue February). https://www.sei.org/publications/land-climate-mitigation-china/

Malins, C., Pereira, L., Popstoyanova, Z., and Riemer, M. 2023. ICF makes big things possible (Issue December). www.icf.com

Marta Camps Arbestain, James E. Amonette, Balwant Singh, T.W. and Schmidt, H.P. 2019. A biochar classification system and associated test methods. *Biochar for Environmental Management* January, 1–2. https://doi.org/10.4324/9780203762264-15

Masud, M.A. Al, Shin, W.S., Sarker, A., Septian, A., Das, K., Deepo, D.M. et al. 2023. A critical review of sustainable application of biochar for green remediation: Research uncertainty and future directions. *Science of the Total Environment* 904. https://doi.org/10.1016/j.scitotenv.2023.166813

Nassary, E.K. 2025. Biochar from maize cob waste: A path to carbon sequestration, enhancing soil health, maize productivity, and climate resilience in Tanzania. *Next Research* 2(2): 100229. https://doi.org/10.1016/j.nexres.2025.100229

Ouro-Salim, O., Guarnieri, P., and Fanho, A.D. 2024. Unlocking value: Circular economy in NGOs' food waste reduction efforts in Brazil and Togo. *Discover Environment* 2(1). https://doi.org/10.1007/s44274-024-00042-4

Owsianiak, M., Lindhjem, H., Cornelissen, G., Hale, S.E., Sørmo, E., and Sparrevik, M. 2021. Environmental and economic impacts of biochar production and agricultural use in six developing and middle-income countries. *Science of the Total Environment* 755. https://doi.org/10.1016/j.scitotenv.2020.142455

Policy: Policies from Central and State Governments. 2018.

Policy – India BioChar and BioResources Network. (n.d.)

Pourhashem, G., Hung, S.Y., Medlock, K.B., and Masiello, C.A. 2019. Policy support for biochar: Review and recommendations. *GCB Bioenergy* 11(2): 364–380. https://doi.org/10.1111/gcbb.12582

Prochnow, F.D., Cavali, M., Dresch, A.P., Belli, I.M., Libardi, N.J., and de Castilhos, A.B.J. 2024. Biochar – From laboratory to industry scale: An overview Brazilian context, and contributions to sustainable development. *Processes* 12: 1006. https://doi.org/10.3390/pr12051006

Promoting Sustainable Agriculture_ASEAN, ERIA, JAIF event addresses crop burning, biochar, and crop insurance progress – ASEAN Main Portal. (n.d.)

Salma, A., Fryda, L., and Djelal, H. 2024. Biochar: A key player in carbon credits and climate mitigation. *Resources* 13(2). https://doi.org/10.3390/resources13020031

Salo, E., Weber, K., Hagner, M., and Näyhä, A. 2024. Nordic perspectives on the emerging biochar business. *Journal of Cleaner Production* 475(September). https://doi.org/10.1016/j.jclepro.2024.143660

Schmidt, H.P. 2013. Switzerland: The first European country to officially approve biochar. *In:* Ithaka. www.ithaka-journal.net/schweiz-bewillingt-pflanzenkohle-zur-bodenverbesserung?lang=en

Scholz, S.M., Sembres, T., Roberts, K., Whitman, T., Wilson, K., and Lehmann, J. 2014. Biochar systems for smallholders in developing countries: Leveraging current knowledge and exploring future potential for climate-smart agriculture. https://doi.org/10.1596/978-0-8213-9525-7

Senadheera, S.S., Withana, P.A., You, S., Tsang, D.C.W., Hwang, S.Y., and Ok, Y.S. 2025. Sustainable biochar: Market development and commercialization to achieve ESG goals. *Renewable and Sustainable Energy Reviews*, 217. https://doi.org/10.1016/j.rser.2025.115744

Siddiqui, S. 2025. Unlocking the environmental potential of biochar: Production, applications, and limitations. *Frontiers in Sustainable Food Systems*, 9. https://doi.org/10.3389/fsufs.2025.1569941

Sörman, L. 2023. Enabling small-scale production of biochar in carbon markets: A multi-actor governance approach. www.kth.se

Štrubelj, L., and Singh Ghaleigh, N. 2025. Biochar regulation in the UK: A wasteful approach to greenhouse gas removal. https://www.research.ed.ac.uk/files/513858803/biochar_waste_regulation_pb.pdf

Sustainability of Biochar Systems in Developing Countries – International Biochar Initiative. (n.d.)

Tisserant, A., Morales, M., Cavalett, O., O'Toole, A., Weldon, S., Rasse, D.P. et al. 2022. Life-cycle assessment to unravel co-benefits and trade-offs of large-scale biochar deployment in Norwegian agriculture. *Resources, Conservation and Recycling* 179. https://doi.org/10.1016/j.resconrec.2021.106030

U.S. Biochar Coalition Response to UNFCCC Article 6 Call for Comments. 2023.

Vuppaladadiyam, A.K., Vuppaladadiyam, S.S.V., Sahoo, A., Murugavelh, S., Anthony, E., Bhashkar, T. et al. 2023. Bio-oil and biochar from the pyrolytic conversion of biomass: A current and future perspective on the trade-off between economic, environmental, and technical indicators. *Science of the Total Environment*, 857. https://doi.org/10.1016/j.scitotenv.2022.159155

Waheed, A., Xu, H., Qiao, X., Aili, A., Yiremaikebayi, Y., Haitao, D. et al. 2025. Biochar in sustainable agriculture and climate mitigation: Mechanisms, challenges, and applications in the circular bioeconomy. *Biomass and Bioenergy* 193. https://doi.org/10.1016/j.biombioe.2024.107531

Waste to Energy Programme | Ministry of New and Renewable Energy | India, 2020. https://mnre.gov.in/en/waste-to-energy/

Yaashikaa, P.R., Kumar, P.S., Varjani, S., and Saravanan, A. 2020. A critical review on the biochar production techniques, characterization, stability and applications for circular bioeconomy. *Biotechnology Reports* 28: e00570. https://doi.org/10.1016/j.btre.2020.e00570

Yu, S., Zhang, W., Dong, X., Wang, F., Yang, W., Liu, C. et al. 2024. A review on recent advances of biochar from agricultural and forestry wastes: Preparation, modification and applications in wastewater treatment. *Journal of Environmental Chemical Engineering* 12(1): 111638. https://doi.org/10.1016/j.jece.2023.111638

Numbered List of Corresponding Links

ERIA: https://www.eria.org/
JIRCAS: https://www.jircas.go.jp/en
NARO: https://www.naro.go.jp/english/
GIZ: https://www.giz.de/en/html/index.html
CABI: https://www.cabi.org/
ASEAN: https://asean.org/

CHAPTER

8

A Comprehensive Review of Biochar Education – Including Formal, Non-Informal and Informal Modalities

Kathleen Draper[1*] and Stephen Joseph[2]

[1] Ithaka Institute for Carbon Intelligence, Canandaigua, NY

[2] University of New South Wales Kensington NSW2052, Griffith University Agricultural Ecology and Technology 170 Kessels Rd, Nathan, QLD 4111

1. Introduction

Education is fundamental to building an effective workforce, building consumer confidence and scaling new innovations and industries. While the need for additional educational materials has been acknowledged by those within the industry for several years, non-profit organizations that support scaling the industry are mostly underfunded with limited resources to develop and deploy educational offerings. Keeping materials up to date is also a challenge given how quickly different sectors of the industry are changing. In addition, there is a growing diversity of stakeholders interested in learning about different aspects of biochar. Determining which stakeholders should be prioritized, what content is most relevant to different audiences and what organizations should be entrusted with developing and deploying science-based, unbiased materials has not been done in a coordinated manner to date.

Stakeholders

A brief description of different stakeholder groups that could benefit from biochar education is provided, along with what content would be helpful and why this information is important to impart to these stakeholders.

Researchers/Academics/Students – The range of academic backgrounds of researchers and students studying biochar has long-since expanded beyond the

*Corresponding author: biocharro2@gmail.com

DOI: 10.1201/9781003467946-11

original focus of agriculture. Today researchers include all types of engineers, material scientists, students in veterinary sciences, building trades, landscape architecture, forestry management and beyond. Amongst the most critical information this group needs is an understanding of how biochars differ, how to test different biochar characteristics, which properties are relevant to different end uses and how to produce and use optimized biochars for different applications.

Biochar producers come from a wide variety of backgrounds such as farming, sewage treatment, waste management, forestry, etc. Very often they have little understanding about biochar production technology options, biochar variability, what local or regional markets might be appropriate for their particular biochar, how to get biochar tested, how to produce and ship it safely and how to navigate and select which carbon market to work with and how to find buyers to tap into carbon removal revenues. Hiring and retaining employees who are knowledgeable about biochar is very challenging currently, so understanding how best to get new employees up to speed on these different topics is increasingly important.

Biochar buyers also come from an increasingly diverse background including composters, urban planners, stormwater management professionals, nurseries, remediation specialists and more recently those in the built environment. As more regulations are enacted pushing for the use of lower embodied carbon products and activities, these professionals are exploring how biochar can be used within their industries. Understanding the properties of relevance for their specific scenarios as well as how and how much to apply for optimal benefit are key areas of focus and will ensure that the best types of biochar are used to meet customer expectations.

The *carbon market ecosystem* is complex and expanding rapidly. Key players in this ecosystem include project developers, standards or protocol developers (e.g. Climate Action Reserve, Carbon Standards International, Puro, Verra), auditors, dMRV providers, brokers or traders and credit buyers. While most of the players have deep experience with other carbon offset or removal strategies, few have much depth of understanding about biochar production and use. Of particular interest to this group is an understanding of what production parameters and end uses impact biochar's durability. In addition, project developers could benefit from best practices related to creating demand for biochar near to production facilities.

Climate activists focused on environmental justice (EJ) generally have a solid understanding of climate change and some of the solutions being deployed to mitigate climate impacts. However, there are some notable situations where the EJ community has hindered the establishment of biochar production facilities by claiming that pyrolysis is similar to incineration in terms of negative environmental impacts or that biochar is toxic. While there may be situations where biochar production facilities may not be advised, it is critical that this community has a solid understanding of the differences between different thermochemical technologies and the variability of biochar production parameters and feedstocks that could potentially produce biochar with higher amounts of toxins so that blanket claims are not made in order to stop construction of biochar facilities.

Additional stakeholders interested in learning more about biochar include waste managers, sustainability managers, policy makers, financiers including investors, bankers, insurers, disaster recovery organizations, renewable energy providers and more.

Types of Education

The number of peer-reviewed papers published in an ever-widening range of publications about biochar has risen sharply from under 5,000 articles per year in 2016 to more than 46,000 in 2023. However, these papers are often behind costly paywalls and/or written using highly scientific language which renders it incomprehensible to non-scientists. The significant growth within the biochar industry over the past few years is hindered by a lack of knowledge at the layman level about biochar – what it is, how it is produced, its benefits, potential uses, variability, biochar standards, properties and what kinds of biochar may qualify for carbon credits. The lack of a nuanced understanding of biochar can lead to purchasing less than optimal biochar for various end uses, which can result in disappointment in meeting customer expectations.

A growing number of formal, non-formal and informal educational options are being developed and deployed by a wide variety of institutions and commercial entities to meet this need.

2. Methods of Education

2.1 Formal

Formal education is normally defined as intentionally organized or structured by institutions, typically colleges or universities, but may also include K-12 education. These offerings are often guided by a curriculum and led by experts or trained professionals.

The first official university level course dedicated to biochar, The Basics of Biochar, has recently been developed by Dr. Oisik Das with support from Luleå University of Technology in Sweden and European Biochar Initiative (EBI) in 2024 (Luleå University of Technology, 2024). The course is targeted at both professionals and students (i.e. Masters or PhD level) that are focused on the environment, biomass utilization, waste valorization and management, biochar production, renewable energy, and/or sustainable materials. After successful completion of the 40-hour on-line course (~USD$400), students receive a certificate from Luleå University of Technology and Biochar Europe as well as 1.5 university credits. A basic knowledge of chemistry and physics is required. The self-study course covers biochar production scenarios, pyrolysis reaction mechanisms, biochar's intrinsic properties, and how biochar can be engineered to enhance the performance properties of materials and structures. The course also includes short video lectures from various biochar experts across the globe.

Biochar lectures embedded into university courses are becoming more common in a growing number of departments such as engineering, landscape design, biomaterials, packaging, renewable energy systems, etc. Some of these are recorded and viewable on YouTube or other channels such as this lecture by Stephen Joseph on the 'Integration of Solar Photovoltaics and Regenerative Agriculture' (Joseph, 2024) at the University of New South Wales in Australia.

Student research projects related to biochar are also proliferating across the globe and are often highlighted at biochar conferences either as a presentation or the information is highlighted in posters for conference goers to view.

Efforts to develop biochar educational materials for K-12 students have been carried out by different academic groups as well as a researcher at NASA. Unfortunately, these materials have not been adopted yet.

2.2 Non-formal

Non-formal education, sometimes referred to as adult or lifelong learning, occurs in an ever-increasing number of places, both virtual and in-person, outside of formal learning institutions. While there is often provided within an organized framework, non-formal education does not necessarily include grading and may or may not result in certification upon completion of a course. Participation in such training is voluntary, and the learning style is often experiential. Examples of non-formal education include community initiated educational offerings such as conferences, professional development courses, study tours, demonstration days, farmer workshops, and webinars.

A growing number of biochar organizations have been developing non-formal educational opportunities which include both in-person and virtual training, study tours, and webinars. These training opportunities are aimed at both beginners of biochar as well as more seasoned professionals in the industry or academia.

In-person

In-person non-formal educational opportunities are flourishing within the biochar industry. These include classes and conferences being offered by different biochar organizations, study tours and demonstration days as well as workshops to show farmers, foresters and others how to make and use biochar using low tech methods. Beyond biochar organizations, these events are provided by individual biochar experts, academics or consultants as well as NGOs working in regenerative agriculture and project developers focusing on biochar as a carbon dioxide removal (CDR) technology.

IBI Biochar Academy – IBI developed the first in-depth, in-person 2-week course to provide a broad overview of biochar, biochar production technologies, carbon markets, techno-economics, market overviews and other key areas of interest within the biochar industry. A shorter one-week version has been developed and deployed in Thailand in May 2024, Kenya in 2025 with others planned in India and other locations. In addition, recordings from the original Biochar Academy are available on-line. New in-person courses under development by IBI include 'Biochar on the Carbon Markets' and 'Business Development for Biochar' to assist emerging biochar entrepreneurs.

Nanjing Agricultural University, University of New South Wales in collaboration with IBI carried out a number of one-week international training courses run by Prof Stephen Joseph from 2013 to 2016. These were based at centers close to operating biochar factories in Jiangsu Province. International experts from around the world came to give lectures to people from all continents. Many of the people who attended these courses went on to establish biochar ventures or to initiate research.

The Australia/New Zealand Biochar Industry Group (ANZBIG) has developed a range of training materials for 1- and 2-day workshops, and these have been held 2 or 3 times a year for the past 7 years. Recently ANZBIG produced a comprehensive guide for farmers and has started producing a manual that provides specific formulations to meet individual soil/plant constraints.

Study Tours

Study tours have typically provided a more in-depth, in-person perspective of the business of biochar production. Pioneered by the Ithaka Institute in 2016 in Nepal, attendees learned all about low-tech biochar production and use in an agricultural setting. IBI hosted its first study tour in Sweden (2017) to highlight biochar production in an urban setting and its extensive use in urban tree planting. Subsequent IBI study tours have been held in Austria (2018), Finland (2019), United States (2023) Switzerland (2023) Italy (2024) and Wales (2025). Those looking to understand the opportunities and challenges of commercial biochar production and liaise with others in the industry can gain valuable insights by visiting existing production facilities. Generally, more than one production facility is included in these tours as well as visits to end-user sites (e.g. farms, concrete manufacturing, etc.). Biochar academics from the region are also invited to share their research. Tours generally last 2–3 days and attract 50–100 attendees. Shorter tours are also increasingly offered as add-ons to biochar conferences.

Conferences and Trade Shows

International, regional, national and sub-national biochar conferences have been hosted since 2009, mostly in-person while some are also available virtually or in hybrid form. Many of the materials from these conferences are made available for free or for a small fee from the host organizations a few months after the event. Some examples of biochar-specific conferences include:

Biochar I, II, III, and IV (Engineering Conferences International) – SIPS is a flagship event of Flogen Stars Outreach organization, a non-profit that hosts all-inclusive science conferences. They began hosting an annual International Symposium on Sustainable Biochar in 2023 in various cities around the world, with Santa Marta, Colombia hosting the event in 2025. This biennial conference is hosted by Engineering Conferences International (ECI), a not-for-profit, global engineering conferences organizer. The event is held at different international locations, with Colombia hosting the 2025 event. The Conference Co-Chairs and Scientific Committee members are made up of biochar experts from around the globe.

Biochar Summit – The Biochar Summit is organized by Biochar Europe (European Biochar Industry Consortium, 2024) for those involved in or interested in the biochar industry including researchers, practitioners, policymakers, and business leaders. The focus on biochar policies and practices in the EU makes this most relevant for those in the European Union.

German Biochar Forum – The German Biochar Forum is an annual event, the largest independent professional biochar event in the German-speaking world organized by the German Biochar Association since 2017. The interactive forum includes expert talks, panel discussions, workshops, poster sessions and an exhibition area for sponsors. The target audience includes academic, commercial, policymakers and all stakeholders along the entire biochar value chain.

The Great Plains Biochar Conference (University of Nebraska Lincoln) Department of Agronomy and Horticulture, the Nebraska Forest Service and the Nebraska Biochar Initiative. This conference is geared towards professionals working in sectors such as research, education, biochar production, agricultural applications, material characterization, organics recycling, soil remediation and carbon sequestration.

Biochar School – The Italian Biochar Association (ICHAR, 2024) has been hosting a two-day, annual event in Italy since 2018. The content includes both academic and commercial material and includes significant information on non-agricultural end-uses for biochar.

North American Biochar Conference – the US Biochar Initiative hosts a biennial conference in different US cities, attracting 600+ attendees. Presentations cover both academic and commercial topics with a focus on helping to build markets for biochar.

The Asia Pacific Biochar Conference (APBC), in collaboration with the International Conference on Biochar Research and Application hosts an annual conference in China. Experts, scholars, and entrepreneurs gather to discuss practical applications, technology trends and promote more sustainable development via pyrolysis, gasification and biochar use.

In Denmark Roskilde University in collaboration with the Energy Cluster Denmark, Food & Bio Cluster, and Clean – the Danish Water and Environment Cluster, hosts an event on pyrolysis innovation within Denmark.

Other conferences and trade shows that have a focus on related topics have begun to include significant biochar content that is of interest to their attendees. For example, the Bioenergy Bioeconomy Event, also known as Bio360 is an annual trade show held in Nantes, France that originally focused on biogas. Over the past few years, the event has expanded to include sustainable transport, biobased materials and Bioenergy with carbon capture and storage or utilization

(BECCUS), renewable gas (gasification, e-methane, bio/circular hydrogen), biochar, solid biomass energy as well as exhibitor abstracts. The event attracts 5,000 attendees from 45 countries. Biochar companies are well represented amongst the 450 exhibiting companies, of whom 42% are international. More than 200 international conference speakers also share information with attendees.

Farmer Workshops & Demonstrations

A growing number of organizations are providing in-person, hands-on workshops focusing on how to make biochar using simple, inexpensive technologies and how to best use biochar with different crops and soils. NGO's such as Warm Hearts Worldwide have taught thousands of smallholder farmers in Africa and Thailand how to turn corn cobs and other agricultural residues into biochar over the past decade. They have recently launched Biochar Life to take over much of this training and help smallholder farmers to tap into the carbon removal marketplace. Other project developers such as PlanBoo, Cotierra, the Carboneers, WongPei and many others are also hosting workshops to train farmers on how to make and use biochar. In addition, other NGOs in the agricultural arena such as Solidaridad are undertaking country-level biochar projects in collaboration with carbon removal market players. After successfully training 120 cotton farmers in Zambia (Solidaridad Network 2025) in early 2024, they plan to roll out training to 100,000 additional farmers across the country within five years.

ANZBIG runs one- and two-day training workshops with Landcare, farmer and conservation groups, agricultural, planning and environmental government departments. The morning sessions are a series of interactive lectures, and the afternoon involves practical exercises. The exercises covered include:

1. Developing specific biochar formulations to meet soil and environmental constraints and plant nutrient requirements
2. Producing a short biochar business plan
3. Developing a marketing strategy for the sale of specific biochar products
4. Designing an on-farm pyrolysis system
5. Producing an engineered biochar in a simple kiln and then applying it to small plots at the training venue
6. Integrating biochar into local waste management programs and projects

ANZBIG has an education and training working group that is developing a syllabus for use in teaching Science, Technology, Engineering and Mathematics subjects

A growing number of demonstration projects are also serving as educational opportunities for different biochar stakeholders. One example is a biochar production demonstration project spearheaded by Cornell University in collaboration with Spruce Haven Dairy and others in Upstate NY with funding provided by the New York State Energy Research and Development agency. They hosted monthly in-person meet-ups geared at showcasing how on-farm biochar production can work for farmers, policymakers and others.

Organizations such as Lost Glacier which has an off-grid location in the French Alps, showcases a micro system that uses wood pellets to produce 24 kWh of electricity, 140 kWh of thermal energy and 5 kg of biochar every day to their guests. They are open year-round and have also hosted an annual event to educate those interested in more sustainable living.

Web-based Training

IBI has offered a technical introductory course for several years, which is helpful to academics or those interested in the deeper science aspects of biochar. More recently, new on-line modules including "All About Biochar" and "What is Biochar?", free micro-learning courses have been developed aimed at educating new entrants to the commercial side of biochar including producers and end-users. Home | IBI Education. Newer offerings include a virtual graduate lecture series for graduate students of universities and an on-line course on 'Biochar on the Carbon Market' for project developers, investors, and carbon credit buyers.

'Getting Going with Biochar' is an online program coordinated by Tejas Rajvihar. The course brings together a cohort of 20 students or graduates from diverse academic backgrounds such as Agriculture, Biotechnology, Food Technology, and Chemical Engineering, among others. These participants are united by their shared interest in biochar—an environmentally significant material with a range of applications, particularly in agriculture and sanitation. The curriculum is designed to provide both theoretical knowledge and practical experience. Students learn from global biochar experts via Zoom presentations followed by Q&A sessions that encourage active engagement and a deeper understanding of biochar science and practice.

Practical Biochar – for Farms, Homesteads and Properties – Regenerative Living in collaboration with Kelpie Wilson developed an introductory program of 9 modules that debuted in January 2025 (Wilson, 2024). The initial live (remote) program was recorded and will be available online. Modules include an Introduction to Biochar, Biochar Chemistry, Making Your Own Biochar, How to Design Kilns and Stoves, Biochar Properties, Biochar for Animals, Biochar and Compost, Biochar in Your Garden or Farm and Healing the Earth with Biochar.

Webinars & Other Online Community Events

A number of different organizations regularly host webinars focused on biochar including the following:

The International Biochar Initiative has hosted webinars for many years on a wide range of topics of interest to the industry which are free for IBI members. Recordings can be accessed on the 'members only' section of their website.

The US Forest Service in partnership with the US Biochar Initiative and the Southern Regional Extension Forestry have been hosting free biochar webinars for the past few years. Webinars generally relate to the production and use of woody biochar from forestry waste. Webinars are recorded but are currently not available through much of the research done by scientists at the United

States Forest Service (USFS) is available on the U.S. Biochar Initiative (USBI) website.

The Green Carbon Webinar is a free series hosted by Christian Wurzer from the UK Biochar Research Centre and European Biochar Research Network (ENCHAR). The series highlights advancements in the field of sustainable carbon materials such as biochar. Each webinar features two 25-minute talks about current research projects or company profiles, followed by live question and answer sessions.

The Minnesota Biochar Initiative hosts a free monthly webinar series called CharTalk featuring various global experts discussing topics such as AI Driven Innovation in Biochar, Webinars are usually announced via MBI's LinkedIn account Minnesota Biochar Initiative: Posts | LinkedIn.

Biochar on Site (previously known as Biochar in the Woods) – A monthly virtual gathering hosted by Kelpie Wilson sponsored by the US Forest Service open to all interested in carbonizing forestry residues.

WhatsApp Groups such as *the Biochar Crusaders* are becoming more popular. This group was established in July 2020 by Mr. R.K. Mehta and friends and now has more than 260 members from India and beyond. Members include biochar manufacturers, biochar equipment manufacturers, producers of biochar amended bio-fertilizers as well as researchers from educational and research institutes.

2.3 Informal Education

Informal education is not pre-planned, it is self-directed with more flexible timing to begin and complete activities. It is less structured and generally not graded nor are certifications or degrees issued upon completion. Examples of informal education are continually expanding across all industries and may include watching educational videos, reading articles, white papers, blogs or books, social media interaction within expert communities, podcasts, hands-on learning through apprenticeships, visiting museums, etc.

Learners can access a vast amount of educational materials at no or low cost, enabling a wider range of individuals the opportunity to gain expertise in biochar.

Books

A growing number of books have been published over the past 15 years aimed at different audiences, from academics to farmers and others interested in understanding biochar at a deeper level. Included below are some of the most relevant books listed in alphabetical order:

The Biochar Debate: Charcoal's Potential to Reverse Climate Change and Build Soil Fertility (Bruges, 2010), paperback. "*The Biochar Debate* was the first book to introduce both the promise and concerns surrounding biochar (fine-grained charcoal used as a soil supplement) to nonspecialists… While explaining the excitement of biochar proponents, Bruges also gives voice to critics who argue that opening biochar production and use to global carbon-credit trading schemes could have disastrous outcomes, especially for the world's poorest people."

The Biochar Handbook: A Practical Guide to Making and Using Bioactivated Charcoal by Kelpie Wilson | Jun 27, 2024; Kindle; paperback (Wilson, 2024). "In accessible and authoritative prose, Wilson demonstrates that biochar is a low-tech but effective means of reducing wildfire risks, restoring soil carbon, managing manure, weaning farms off of toxic inputs, and producing the best compost ever made." The book includes instructions for making biochar, how to use it in soils to boost water retention and recipes for potting soils and more.

Biochar for Environmental Management: Science, Technology and Implementation, 3rd Edition by Lehmann and Joseph (2024). "This book not only reviews recent advances made in our understanding of biochar properties, behavior, and effects in agriculture, environmental management, and material production but specifically develops fundamental principles and frameworks of biochar science and application. This third edition has been fully revised and updated to reflect recent developments and growing trends, with important coverage of the application of biochar outside of its traditional soil-based uses, the commercialization of biochar, and its incorporation into policy."

Gardening with Biochar: Supercharge Your Soil with Bioactivated Charcoal: Grow Healthier Plants, Create Nutrient-Rich Soil, and Increase Your Harvest (Cox 2019). In this highly accessible handbook, long-time garden writer Jeff Cox explains what biochar is and provides detailed instructions for how it can be made from wood or other kinds of plant material, along with specific guidelines for using it to enrich soil, prevent erosion, and enhance plant growth.

The Biochar Revolution: Transforming Agriculture & Environment (paperback) by Taylor and McLaughlin (2010). Book chapters cover: Ancient Solutions to Modern Problems, How to Produce Biochar, How to Test and Use Biochar, How to Change the World with Biochar, The Financial Benefits of Biochar, and How to Garden and Farm with Biochar.

The Biochar Solution: Carbon Farming and Climate Change by Albert K. Bates (Bates, 2010). This book provides a brief history of the discovery of terra-preta, agricultural practices climate impacts, and how carbon can be stored in soils to improve soils and rebalance carbon.

Burn: Using Fire to Cool the Earth (Bates, Draper and Rider 2018) hard cover, paperback, kindle, audio. "Burn looks beyond renewable biomass or carbon capture energy systems to offer a bigger and bolder vision for the next phase of human progress, moving carbon from wasted sources: into soils and agricultural systems to rebalance the carbon, nitrogen, and related cycles; enhance nutrient density in food; rebuild topsoil; and condition urban and agricultural lands to withstand flooding and drought; to cleanse water by carbon filtration and trophic cascades within the world's rivers, oceans, and wetlands; to shift urban infrastructures such as buildings, roads, bridges, and ports, incorporating drawdown materials and components, replacing steel, concrete, polymers, and composites with biological carbon and to drive economic reorganization by incentivizing carbon drawdown."

The Farmer's Guide (ANZ Biochar Industry Group) is an e-book ($35AUS) and hard copy via Amazon ($85 AUS) that compiles the experiences of gardeners, farmers, and researchers who have made and used biochar to produce food crops in home gardens, smallholdings, and farms around the world. It draws on the collective wisdom of people who have used biochar for centuries, on innovations developed in the last 100 years, and on the latest scientific research. The 11 chapters cover: Principles of biomass pyrolysis for biochar production; Indigenous practices of biochar production and utilization; Choosing a biochar unit that meets your needs; Properties of fresh biochars and wood vinegar; Properties of biochar change after interaction with soils, compost, or animals; Effects of biochar on crop production and soil properties; How to choose and enhance biochars to meet specific soil constraints; Methods for applying biochar; Case studies; Biochar in regenerative agriculture and carbon drawdown and Testing biochars.

Making Charcoal and Biochar: A Comprehensive Guide, paperback by Oaks, (2018) Kindle; Paperback. "This book gives a wide range of possibilities for making charcoal on a small scale and for commercial production. There are chapters on the heritage skills of earth burns, the enduring popularity of metal kilns, and the future represented by the charcoal retort. Biochar – or small particle charcoal – has been heralded as an ancient but rediscovered 'super substance' that can increase soil fertility and productivity whilst locking up carbon into the ground. This book looks at the ongoing discussion and weighs up the evidence. It concludes with a celebration of the myriad ways in which charcoal can be put to use."

Sacred Soil: Biochar and the Regeneration of the Earth by Tindall, Apffel-Marglin and Shearer (2017); kindle, paperback. This book describes "how utilizing the biochar embedded in terra preta, the recently rediscovered sacred soil of the pre-Columbian peoples of the Amazon rainforest, can cut our dependency on petrochemicals, restore the health of our soils, remove carbon from our overheating atmosphere, and restore the planet to pre-industrial levels of atmospheric carbon by 2050."

Terra Preta: How the World's Most Fertile Soil Can Help Reverse Climate Change and Reduce World Hunger, paperback by Scheub et al. (2016). The authors "condensed all the knowledge about the world's most fertile soil into a convenient guidebook. In addition to a sound instruction manual on producing terra preta and organic charcoal (biochar), the handbook covers fundamental principles from climate farming to closed-loop economy. It makes a passionate plea against synthetic fertilizers and genetic technology."

Fact Sheets and White Papers

A number of short, topic-specific fact sheets have been developed by USBI in an effort to provide unbiased information for biochar producers and end users. Topics include Biochar application, use in compost, how to interpret lab results and many more. Fact sheet | US Biochar Initiative.

Lengthier papers have been written describing how biochar production and use could help decarbonize and improve different industries and cropping systems. White papers for the Coffee Industry (IBI n.d.), Maize Producers (Draper 2018) and the Dairy industry (Draper and Sweet 2020) have been sponsored by IBI and the Innovation Center for US Dairy.

A white paper that provides costs and benefits from users of biochar has been produced by ANZBIG 'A report on the value of Biochar and wood vinegar' (Robb and Joseph 2020).

Podcasts

The topic of biochar has been featured in a wide variety of podcasts such as Nori, Roots and All, No Till Farmer, The Permaculture Podcast, The Gardenerd, Voices from the Field and many more. There are a few podcasts that are largely or completely dedicated to biochar including:

The Biochar Podcast - 11 episodes covering a wide variety of end use and other biochar topics from 2021 to 2023 produced by ANZBIG.

Grounded: A Climate Startup Journey is a podcast about setting up a new biochar production facility in the United Kingdom. Host Tom Previte, who also hosts The Carbon Removal Show podcast, captures the highs and lows of a biochar startup – from sharp learning curves to unlikely breakthroughs he highlights the realities and complexities of setting up a carbon removal project, and gets to grips with the science, the money, and everything it takes to build an environmentally (and financially) sustainable business.

Carbon Conversations is hosted by Biochar.Life. Conversations with biochar experts. Started 2023, the podcast highlights experts pioneering innovative environmental solutions as well as business leaders championing sustainability in their sectors. From the intricacies of biochar and renewable energy to the broader scopes of smart agriculture, Carbon Conversation's aim is to shed light on the multifaceted approaches to addressing the planet's pressing challenges.

IBI – Biochar Production Technologies (Biochar Technology Podcast n.d.) Researcher Akio Enders hosted an introduction to biochar production technology podcast as well as interviews with various biochar production technology manufacturers in 2022.

On-line Forums

The Biochar Discussion Forum was established in 2006 by Tom Miles, Ron Larson and Eric Knight as a mechanism to encourage conversations amongst newcomers and established biochar practitioners about the production and use of biochar and the biochar industry. More than 1,600 active members share knowledge, exchange ideas, highlight resources, and work to further the development of biochar. The 'list' discourages promoting any one specific corporation, but product information and factual information about biochar manufacture or application is encouraged. This discussion is curated to maintain high quality interaction amongst professionals actively collaborating to help build the industry. An archive of over 40,000 posts dates back to 2010.

LinkedIn and FaceBook are also good sources for information about biochar though some of the information may be of a more promotional nature versus science based.

Websites

Websites contain information about biochar abound as the number of biochar companies and organizations supporting it continues to grow. Many of these are company websites that have science-based educational information aimed at providing relevant information to potential buyers. International, regional, national and sub-national biochar association websites are also often good sources of educational materials, some of these include:

The Australia/New Zealand Biochar Industry Group – The ANZBIG website contains a repository of webinar recordings, newsletters and a link to download the Farmer's Guide to Biochar.

The Biochar Atlas was spearheaded by the US Department of Agriculture in collaboration with the US Department of the Interior (Pacific Northwest Biochar Atlas n.d.). This toolkit which helps identify soils that could most benefit from biochar originally focused on the Pacific NorthWest but it is being expanded to include a national focus with enhanced functionality.

The Biochar Journal – Contains a variety of science-based articles about biochar that are practical in nature with the intent of inspiring others to make, test and use biochar in different ways.

Biochar Today has become a go-to resource for summaries of recent peer reviewed research as well as current news within the biochar industry and blog interviews with various biochar industry experts.

International Biochar Initiative – IBI's website contains various research summaries and information papers, a biochar classification tool and information about different on-line and in-person educational opportunities. Educational content from the Biochar for Sustainable Soils pro Biochar for Sustainable Soils (B4SS) project, a United Nations Environment Programme/Global Environment Facility (UNEP/GEF) funded project that established 13 demonstration sites that compared biochar with alternative practices in six countries over four years. The internal reporting to UNEP/GEF collated lessons learned for small-scale farmers about the production and use of biochar in different cropping systems. A list of current biochar associations is also available (IBI n.d.) as is a new job portal for biochar related jobs (IBI n.d.).

The Nordic Biochar Network website has a repository of webinar recordings, a list of upcoming biochar events and a map of academic and commercial biochar activities in the region.

US Biochar Initiative – USBI's website contains an extensive learning repository that includes case studies, fact sheets on various biochar topics, recorded webinars and presentations, newsletters, etc.

3. Discussion

Biochar education, both formal and informal methods, has been on the rise over the past five years. In-person educational opportunities include conferences, study tours and week-long biochar academies. Online opportunities include a growing number of webinars, websites and on-line forums. Most educational materials are developed and deployed by national, regional or international biochar NGOs with limited funding. To date much of the funding has come from philanthropic organizations and biochar companies. In Australia state and local government, universities and NGO's have all provided funding for training workshops and the development of training material. The U.S. Forest Service has also provided significant funding for education related to biochar in the U.S. Training in the Global South is needed but the ability to pay for such training is limited making deployment to those in Africa, LatAm and Asia a challenge.

Navigating the biochar education landscape to find the most relevant, up-to-date information is a challenge. Educating the growing number and diversity of new biochar stakeholders is therefore also difficult. Resources, both financial and human, to develop and deploy biochar education remain scarce. Determining which stakeholders prioritize will vary based on the perspective of the various actors within the biochar eco-system and many aspects of the industry are changing rapidly causing educational materials to become out of date quickly (e.g. standards, carbon market protocols, policies and business models). Streamlining and collaborating on the development of educational materials across borders could help make education more cost-effective.

4. Recommendations

There are several recommendations that could lead to increased educational outreach and ultimately help scale the industry including those discussed below.

Customizing biochar education to local or regional needs is important, but there are benefits to having a centralized, reliable, up-to-date repository of educational materials that could be translated and customized by others. Creating a networked community of 'Char Hubs' that collaborates on identifying the most important educational materials and develop and deliver these at a local level.

While IBI does an excellent job at tracking all new peer reviewed papers in a monthly service for members, most academic papers are expensive to access in full and difficult to interpret for non-scientists. There are also too many new peer reviewed papers coming out each month to be able to dive deep into each paper. Creating some type of searchable database with AI generated summaries in plain language with the ability to generate meta-analyses on a wide variety of biochar topics would provide significant value and confidence to both producers and users of biochar.

Embedding biochar education materials in curriculum for K-12 (e.g. science, math, history) and more University/College degree programs including agriculture, climate, engineering, environmental studies, forestry, architecture, landscape architecture, renewable energy, veterinary and more would explore vast numbers of younger people to biochar.

Developing Continuing education (CE) courses for those licensed trades that require professionals to take courses to maintain their licenses (e.g. engineering, architecture, construction, etc.) would incentivize professionals in these trades to learn how biochar could help reduce embodied carbon.

Developing courses aimed at educating different governmental bodies involved in energy, climate, environmental protection (EPA, soil & water conservation, etc.), agriculture, forestry could inspire new policy and regulations that could help scale the industry.

Developing courses aimed at educating agricultural extension agents, NGOs focused on regenerative agriculture, farmer cooperatives, would provide a train-the-trainer capability within a high priority market.

Case studies for biochar use in different agriculture cropping scenarios as well as other end use markets should be developed to help educate end users about the financial benefits to using biochar. Case studies about biochar production scenarios based on real life demonstration plants could help increase investor confidence in supporting biochar production.

Training provided to large global food and beverage companies that are focused on reducing scope 3 emissions on farms, could aid biochar adoption by such companies. In addition, education focusing on sectors such as wastewater management, the construction and demolition industry and others managing large amounts of waste could boost biochar scaling in these industries.

5. Conclusions

In this chapter a review of the increasing variety of stakeholders in need of different types of biochar knowledge was presented including academics, biochar producers and buyers, those in the carbon removal markets ecosystems and more general climate activists. A variety of methods and opportunities for biochar education were outlined as well. Education has long been seen as a barrier to scaling the biochar industry. Many different NGOs and other entities have stepped up to try to fill this gap in recent years with minimal funding.

As new industries become interested in biochar, new educational materials that address the specific needs and biochar opportunities are needed. Given the ever-changing dynamics of the carbon removal marketplace, it is critical for this type of information to also be continually updated for producers and buyers alike. Moving forward it is important to identify ways to collaborate across different geographies to ensure that consistent and up-to-date information is provided to as many stakeholders as possible.

Acknowledgments

Valeria Araico, Albert Bates, Luisa Marin, Tom Miles, Tejas Rajvihar,

References

ANZ Biochar Industry Group. (n.d.). *A Farmer's Guide: To the Production, Use and Application of Biochar*. https://www.anzbig.org/farmers-guide.html

Australia New Zealand Biochar Industry Group. (n.d.). *Australia New Zealand Biochar Industry Group (ANZBIG)*. Retrieved [insert date], from https://www.anzbig.org/

Bates, A.K. 2010. *The Biochar Solution: Carbon Farming and Climate Change*. New Society Publishers.

Bates, A., and Draper, K. 2019. *Burn: Using Fire to Cool the Earth*. Chelsea Green Publishing.

Bio360 Expo & Conference. (n.d.). *Bio360 Expo & Conference*. https://www.bio360expo.com/

Biochar Discussion Forum. (n.d.). *main@Biochar.groups.io | Home*. Retrieved [insert date], from https://biochar.groups.io/g/main

Biochar on Site. (n.d.). *Biochar On Site – Stewardship Biochar for Social Impact*. https://biocharonsite.org/

Biochar Technology Podcast [Audio podcast]. YouTube. https://www.youtube.com/watch?v=W4oKZQezrYE

Biochar Today. (n.d.). *Home*. Retrieved [insert date], from https://biochartoday.com/

Biochar US. (n.d.). *Home*. Retrieved [insert date], from https://biochar-us.org/

Bruges, J. 2010. *The Biochar Debate: Charcoal's Potential to Reverse Climate Change and Build Soil Fertility* (Schumacher Briefing No. 16). Chelsea Green Publishing.

Carbon Conversations. (n.d.). [Audio podcast]. Spotify. https://open.spotify.com/show/7nQsF4TlW9byuMTyP1xst1

Carboneers. (2024). *Monitoring, Reporting & Verification (MRV)*. https://www.carboneers.earth/mrv/

Cotierra. (n.d.). *Cotierra* [Website]. https://www.cotierra.com/

Cox, J. 2019. *Gardening with Biochar: Supercharge your Soil with Bioactivated Charcoal—Grow Healthier Plants, Create NutrientRich Soil, and Increase your Harvest*. Storey Publishing.

Draper, K. 2018, July. *The Potential for Biochar to Deliver Greater Sustainability for Maize Cultivation and Processing: A White Paper.* International Biochar Initiative & Ithaka Institute for Carbon Intelligence. https://biochar-international.org/wp-content/uploads/2018/11/IBI-Maize-White-Paper-2018.pdf

Draper, K., and Sweet, S. 2020. *The Potential for Biochar to Enhance Sustainability in the Dairy Industry: A Synthesis Report* (Innovation Center for U.S. Dairy). Ithaka Institute for Carbon Intelligence & Cornell University. https://cpb-us-e1.wpmucdn.com/

blogs.cornell.edu/dist/c/8159/files/2020/09/DairyManureBiocharSynthesisReport_InnovationCenter_2020.pdf

Engineering Conferences International. (n.d.). *Bio-char IV*. Engineering Conferences International. https://engconf.us/conferences/materials-science-including-nanotechnology/bio-char-iv/

European Biochar Industry Consortium. 2024, July 15. *SAVE THE DATE: The Biochar Summit 2025*. https://old.biochar-industry.com/2024/save-the-date-the-biochar-summit-2025/

German Biochar Forum. (n.d.). *German Biochar Forum*. https://german-biochar-forum.com/en

Green Carbon Webinar. (n.d.). *Green Carbon Webinar* [Website]. https://www.greencarbonwebinar.org/

Grounded: A Climate Startup Journey. [Audio podcast]. Spotify. https://open.spotify.com/show/3jL31BF1zhbV6V1cF9nyEn

International Biochar Initiative. (n.d.). *Biochar Education*. Retrieved from https://learn.biochar-international.org/

International Biochar Initiative. (n.d.). *Biochar Jobs Portal*. Retrieved from https://biochar-international.org/resources/biochar-jobs-portal/

International Biochar Initiative. (n.d.). *Home*. Retrieved from https://biochar-international.org/

International Biochar Initiative. (n.d.). *Regional Biochar Groups*. Retrieved from https://biochar-international.org/resources/regional-biochar-groups/

International Biochar Initiative. (n.d.). *The Potential for Biochar to Deliver Greater Sustainability for Coffee Cultivation and Processing*. https://biochar-international.org/the-potential-for-biochar-to-deliver-greater-sustainability-for-coffee-cultivation-and-processing/

Italian Biochar Association (IC HAR). 2024, July. *Programma 8th ICHAR Biochar School* [PDF]. https://ichar.org/wp-content/uploads/2024/07/V5-programma-8th-ICHAR-Biochar-School-.pdf

Joseph, S. 2024. Integration of solar photovoltaic, regenerative agriculture. UNSW School of Photovoltaic & Renewable Energy Engineering. https://www.youtube.com/watch?v=e3zPkgQUR3I

Lehmann, J., and Joseph, S. 2024. *Biochar for Environmental Management: Science, Technology and Implementation* (3rd ed.). Routledge.

Lost Glacier. (n.d.). *Lost Glacier* [Website]. https://www.lostglacier.org/

Luleå University of Technology. (n.d.) 2024. The basics of biochar – Uppdragsutbildning. Luleå University of Technology. https://www.ltu.se

Minnesota Biochar Initiative. (n.d.). *Posts* [LinkedIn page]. Retrieved [insert retrieval date], from https://www.linkedin.com/company/minnesota-biochar-initiative/posts/?feedView=all

Nordic Biochar. (n.d.). *Home*. Retrieved from https://www.nordicbiochar.org/

Oaks, R. 2018. *Making Charcoal and Biochar: A Comprehensive Guide*. The Crowood Press.

On-line forums

Pacific Northwest Biochar Atlas. (n.d.). *Home*. Retrieved from https://www.pnwbiochar.org/

Planboo. (n.d.). *Planboo* [Website]. https://planboo.eco/

Planboo. (n.d.). *WongPhai Biochar Project* [Webpage]. https://planboo.eco/project/wongphai-biochar-project/

Podcasts Regenerative Living. (n.d.). *Courses*. Retrieved from https://www.regenerativeliving.online/courses

Robb, S., and Joseph, S. 2020, April. *A Report on the Value of Biochar and Wood Vinegar: Practical Experience of Users in Australia and New Zealand* (Version 1.2). Australia New Zealand Biochar Industry Group. https://anzbig.org/wp-content/uploads/2020/07/ANZBI-2020-A-Report-on-the-Value-of-Biochar-and-Wood-Vinegar-v-1.2.pdf

Scheub, U., Pieplow, H., Schmidt, H.-P., and Draper, K. 2016. *Terra Preta: How the World's most Fertile Soil can Help Reverse Climate Change and Reduce World Hunger.* Greystone Books.

Solidaridad Network. (2025, January 13). *From Waste to Wonder: Zambian Farmers Embrace Biochar in New Partnership*. https://www.solidaridadnetwork.org/news/from-waste-to-wonder-zambian-farmers-embrace-biochar-in-new-partnership/

Solidaridad Network. (n.d.). *Home*. https://www.solidaridadnetwork.org/

Taylor, P., and McLaughlin, H. 2010. *The Biochar Revolution: Transforming Agriculture & Environment*. Global Publishing Group.

The Biochar Journal. (n.d.). *Home*. Retrieved from https://www.biochar-journal.org/en/home/

The Biochar Podcast. (n.d.). *The Biochar Podcast* [Audio podcast]. Spotify. https://open.spotify.com/show/3xCcSif12wCM3piXygkJqU

Tindall, R., ApffelMarglin, F., and Shearer, D. 2017. *Sacred Soil: Biochar and the Regeneration of the Earth.* North Atlantic Books.

University of Nebraska–Lincoln. (n.d.). *Great Plains Biochar Conference*. Department of Agronomy & Horticulture. https://agronomy.unl.edu/great-plains-biochar-conference/

US Biochar Initiative. (n.d.). *Fact Sheet*. https://biochar-us.org/type-resource/fact-sheet

Wilson, K. 2024. *The Biochar Handbook: A Practical Guide to Making and Using Bioactivated Charcoal* (pp. 288). Chelsea Green Publishing.

CHAPTER

9

Biochar Unveiled: Real-World Applications and Visionary Insights for the Developing World

Sara de Jesus Duarte[1]* and Thomas Miles[2]

[1] The Next 150, Mexico City, México

[2] T R Miles, Technical Consultants, Inc. Portland, Oregon, USA

1. Introduction

The story of biochar has evolved from a singular agricultural solution to a revolutionary, multi-sector material that addresses some of humanity's most pressing challenges. What began as a soil amendment technology has transformed into a versatile carbon-based material with applications spanning from climate mitigation to advanced manufacturing, from water purification to pharmaceutical development. This transformation represents more than technological advancement, it embodies a paradigm shift toward sustainable solutions that can simultaneously address environmental, economic, and social challenges in the developing world.

The expansion of biochar beyond agriculture began earnestly around 2009, gained momentum through the 2010s with breakthrough applications in water treatment, and has experienced explosive growth since 2018 across multiple industrial sectors. Today, biochar serves roles in energy storage, pharmaceutical manufacturing, construction materials, and environmental remediation demonstrating its unique position as a sustainable material capable of addressing multiple global challenges simultaneously.

This multi-sector expansion has been accompanied by remarkable market growth. The global biochar market, valued at USD 600 million in 2023, with a remarkable 91% Compound Annual Growth Rate (CAGR) production growth rate from 2021 to 2023, is projected to reach USD 3.3 billion by 2025 (IBI 2023;

*Corresponding author: saraduarteufrb@gmail.com

DOI: 10.1201/9781003467946-12

Earth 2024). This growth trajectory reflects not only increasing market demand but also the material's proven versatility and effectiveness across diverse applications.

The significance of biochar's versatility becomes particularly pronounced when viewed through the lens of developing world challenges. In regions where resources are limited, and environmental pressures are mounting, a single material that can improve agricultural productivity, provide clean water solutions, generate renewable energy, and contribute to sustainable construction offers unprecedented opportunities for integrated development strategies. This chapter underscores the importance of biochar in this context, drawing from experiences in both developed and developing countries to support scalable solutions and initiatives tailored to emerging economies.

The chapter is divided into four parts: Part I examines Primary Production Sectors biochar's application in agriculture and forest, Part II covers industrial uses, Part III addresses environmental services, and Part IV explores emerging applications.

Each section combines real-world implementation with forward-looking insights from researchers, practitioners, and industry leaders. It not only showcases current achievements but also addresses ongoing challenges, emerging solutions, and future directions. Technical guidance is also provided, including best practices for feedstock selection, pyrolysis conditions, such as temperature and retention times, to help bridge the gap between theory and practical application.

Through this comprehensive approach, the chapter demonstrates that biochars are more than a collection of isolated applications—they are transformative materials capable of supporting integrated, sustainable development strategies uniquely suited to the needs and opportunities of the developing world.

2. Methodology

Interviews were conducted with experienced researchers and CEOs with extensive backgrounds in biochar applications across multiple sectors. These industry experts represent diverse fields including primary production sectors like agriculture and forestry, industrial applications encompassing material sourcing, energy generation, and construction, environmental services covering water treatment and decontamination, and emerging applications in renewable energy, pharmaceutical potential uses, and animal feed.

Each interview explored current biochar applications within the participant's sector alongside their projections for future market development. The discussion framework addressed several key areas to ensure comprehensive coverage while maintaining comparability across different industries. Participants provided detailed project overviews describing their current biochar initiatives and operational scope. They identified specific challenges and market problems that biochar technology addresses within their respective sectors, from soil degradation in agriculture to contamination issues in environmental remediation.

Technical aspects received significant attention, with experts discussing their feedstock selection criteria, production methodologies, and quality control measures. They shared experiences that validate biochar's effectiveness in their applications. Barriers to implementation emerged as a critical discussion point, covering technical limitations, economic constraints, regulatory hurdles, and market adoption challenges that projects encounter during development and scaling phases.

Participants shared long-term projections for biochar integration and potential market transformation within their industries. The conversations also explored innovation pipelines, including emerging applications, novel product development opportunities, and previously unexplored market segments. Finally, sector-specific requirements were examined, addressing industry standards, regulatory compliance needs, and specialized application parameters that influence biochar adoption and commercial viability across different markets.

3. Part I: Primary Production Sectors: Agriculture and Forest

3.1 Agricultural Applications

Biochar is a transformative soil amendment that addresses agricultural challenges while providing environmental benefits. Applications typically increase yields by 10–30% in degraded soil through improved nutrient retention, pH correction, and enhanced water management (Jeffery et al. 2011; He et al. 2017). The material's high surface area (200–400 m^2/g) and alkaline nature (pH 8–10) optimize soil conditions for plant growth and microbial activity (Liu et al. 2017; de Jesus Duarte et al. 2019).

Biochar significantly improves water retention, particularly in sandy soils where field capacity increases by 23.8% and available water capacity by 25.6% (Bruun et al. 2014). These water management benefits are increasingly valuable given climate variability challenges. Additionally, biochar can reduce fertilizer requirements by 10–30% (Jia et al. 2021; Ye, et al. 2020) while sequestering approximately 1 to 2.5 metric tons of CO_2 equivalent tons per metric ton produced (Stavrakou et al. 2022). See chapter 5 of this book, for crafting a potent product for the carbon market. See chapter 2 of this book to create a potent biochar soil amendment.

Commercial Implementation: Sonnenerde GmbH Case Study

Austrian company Sonnenerde GmbH[1] exemplifies successful biochar commercialization, expanding from 150 tons annual production capacity in 2012 to 1,000 tons by 2023. The company's evolution demonstrates key success

[1] https://www.sonnenerde.at/de/

factors for biochar adoption. Their feedstock optimization process involves combining feedstocks with different properties for pyrolysis, screening residues from composting, cleaning the biomass through water baths and enriching minerals during processing. Sonnenerde's product diversification strategy addresses multiple market segments, from home gardeners using "Organic Black Soil" products to specialized agricultural applications including feed chars and manure chars for treating digestate.

Their success highlights two critical implementation principles: biochar should never be applied directly to fields but must be enriched with nutrients and microorganisms, and applications should focus on concentrated spot treatments rather than broadcast application to maximize impact while managing costs.

The company's experience reveals ongoing economic challenges, with biochar costs in developed markets ranging from $80–100 per cubic yard and application costs reaching $430–3,500 per acre. However, Dominik Dunst from Sonnenerde identifies water retention capacity as biochar's most valuable agricultural property, particularly as climate variability increases.

Carbon Remove

In Brazil, a farmer-led company was founded after achieving a 30% increase in livestock stocking rates by applying biochar. According to founder and producer Guilherme Marques Carbon Remove[2] now offers a full-service "biochar as a service" model for farmers and agro-industries, managing everything from pyrolysis to carbon credit generation—either on farms or at industrial sites. Targeting operations with residual biomass and available land, they focus on common feedstocks like sugarcane bagasse, forestry residues, coffee husks, and rice husks. Currently, they are negotiating industrial-scale projects to build a low-carbon value chain in agriculture. This nature-based approach supports soil health and could drive a regenerative agriculture revolution. Another successful example in Brazil is Net Zero[3], discussed in Chapter 3.

Biochar Applications in Sustainable Agriculture: Circular Bio-nutrient Economy Network (CBEN)

The Circular Bionutrient Economy Network (CBEN)[4]—represented by Erick Abala, Bella Coulotta, and Rita Ellen—is a multi-sector network advancing sustainable agriculture by promoting nutrient recovery from organic waste such as crop residues, market waste, and excreta. With over 100 members from 50+ organizations across 14 countries (mainly in Sub-Saharan Africa), CBEN

---

2 https://www.carbonremove.com.br/

3 https://netzero.green/en/biochar/

4 https://www.cbenetworks.org/post/unlocking-the-power-of-biochar-innovations-and-collaboration-in-african-agriculture

connects researchers, policymakers, farmers, and entrepreneurs to support bio-nutrient circularity.

CBEN works directly with farmers to produce biochar, organizing artisanal production tours, regional conferences, and virtual working groups to build confidence, share knowledge, and promote scalable solutions. The network sees biochar as a key tool for enhancing food security in Africa, where degraded soils, organic waste availability, and urgent fertilizer need create ideal conditions for impact. Kenya is emerging as a regional leader due to its venture investment activity, digital tools like DMRVs, and engagement in climate policy.

Biochar's ability to improve soil health, boost fertilizer efficiency, and sequester carbon aligns with Africa's growing participation in carbon markets. To accelerate adoption, CBEN recommends:

- Localizing narratives to reflect farmers' realities
- Investing in innovation, such as Carbon Standards International (CSI)[5] certified laboratory
- Developing policy incentives that link biochar to national strategies
- Building regional knowledge hubs for training and farm networks
- Quantifying impacts through success stories and data from smallholders and producers

By leveraging local resources, cross-sector collaboration, and inclusive policy frameworks, CBEN positions biochar as a regenerative, climate-resilient solution for restoring soils and securing Africa's agricultural future.

3.2 Forestry Applications

Forest residue utilization for biochar production addresses multiple environmental challenges simultaneously. Converting forest residues that would otherwise remain on-site into biochar significantly reduces wildfire risk while eliminating high costs associated with wildfire suppression. This approach transforms waste materials into valuable soil amendments while protecting rural communities and reducing landslide risks.

USDA Forest Service Research Program Case Study

Dr. Carlos Rodríguez-Franco's research at the USDA Forest Service[6] demonstrates biochar's effectiveness in forest ecosystem restoration. Use of mobile biochar production technologies, including kilns capable of reaching 600°C, minimizes transportation costs while processing forest residues on-site.

---

[5] https://www.carbon-standards.com/en/standards/service-514~production-of-biochar.html

[6] https://www.fs.usda.gov/

The research program's focus on abandoned mine land restoration showcases biochar's versatility. Woody biochar effectively retains heavy metals including lead, mercury, and cadmium, while its key properties—pH, surface area, and water retention—support ecosystem restoration efforts. For every 10 tons of biochar applied, significant improvements in soil carbon content and carbon sequestration result.

Implementation Challenges and Solutions

Despite proven benefits, biochar adoption in forestry faces several barriers. These include limited awareness, public misconceptions about environmental impacts, inconsistent policies, and absence of universally accepted standards. The international market for forestry biochar applications remains underdeveloped, requiring strategic approaches to market development.

Dr. Rodríguez-Franco's strategy emphasizes stakeholder education, building partnerships to leverage complementary strengths, and collaboration with agencies to promote the best management practices. His work demonstrates that successful biochar implementation requires addressing competing uses for residues like activated carbon and wood chips for paper and solid wood products while developing clear value propositions for specific applications.

Economic and Policy Considerations

The biochar industry's maturation requires supportive policy frameworks, particularly carbon pricing mechanisms and investment security. Experts recommend government-secured carbon removal prices to provide stable, long-term funding necessary for significant capital investments in pyrolysis technologies. Additional government support during investment phases would accelerate industry development and cost reduction through facility scaling.

Success in both agricultural and forestry applications demands more than technical proficiency, requires market intelligence, regulatory compliance, innovative product development, distribution, and strategic value creation. The experiences of companies like Sonnenerde and research programs like those led by Dr. Rodríguez-Franco provide valuable roadmaps for scaling sustainable soil management practices globally, demonstrating that biochar adoption succeeds through accumulating real-world successes across diverse applications and market segments.

Future Applications in Agriculture and Forest

To expand the biochar's role in current and future agricultural practices, its production costs need to be reduced. Water holding capacity are among the most important properties of biochar for agriculture. Additionally, scaling up biochar's use requires investments in resources and changes, as well as investment security. For example, securing a carbon removal price set by governments would make it easier to calculate the real cost of biochar production. Further support from governments is also needed to develop pyrolysis technologies during the investment phase.

A key strategy for future agricultural applications involves comprehensive product development, incorporating biochars into diverse agricultural solutions including geotechnical improvements, chemical fertilizer compositions, organic fertilizer formulations, and as an ingredient in various substrate types, which will be essential for market advancement.

Biochar applications are expected to expand into several key areas including carbon sequestration through enhanced focus on forest-based carbon credit programs, integrated forestry systems that combine biochar production with sustainable forest management practices, and bioenergy co-production that generates both biochar and renewable energy. While the sector faces challenges in scaling production and standardizing quality, technological advances and increasing environmental awareness are driving continued innovation and adoption in forest-related applications.

4. Part II: Industrial Applications

Biochar has emerged as a versatile material for diverse industrial applications. It offers solutions for sustainability, waste management, and carbon sequestration. Additionally, it addresses industrial needs such as substitutes for fossil fuels in steelmaking and pollution reduction in cement, reduction of industrial odors and by being a key ingredient to filter contaminants in water/air. A detailed description of those applications can be found in chapter 4 of this book.

In this topic, we interviewed producers who make biochar as an ingredient for industrial applications such as biochar a material source for industry, water filtration, construction materials, and carbon composites. Future applications include steel production, electronics, and energy storage. The needs addressed by biochar include carbon reduction, replacing fossil fuels in cement and steel production, converting forestry and agricultural residues to higher value products, filtering air and water, reducing industrial odors, and providing synthesis gases to supply renewable energy as gases or potential liquid fuels.

4.1 Material Source for Industry

Industrial production utilizes slow pyrolysis at 300–800°C with 20–30-minute residence times. Autothermal pyrolysis and gasification at 500–900°C optimize carbon content and surface area. Extended processing beyond two hours enhances stability for construction and steel applications (Brown and del Campo 2015). Industrial systems generate substantial high-temperature char while providing energy for biomass industries (Crespo-Barreiro et al. 2023; Huang et al. 2020; Ibitoye et al. 2024).

Applications demonstrate significant environmental benefits: 20% greenhouse gas reduction in concrete blends and 19–28% emission reductions when substituting 20% coke with biochar in steel production (Lyu et al. 2022; Roberts et al. 2023; Woolf et al. 2010). Despite challenges including production costs and scalability limitations, demonstration projects by Novocarbo[7], Wakefield Biochar[8], and Made of Air GmbH[9] showcase transformative potential across diverse industrial applications.

Industry Leadership and Innovation Case Study

Made of Air GmbH (MOA) showcases successful biochar industrialization by developing carbon-negative materials across sectors. MOA has advanced innovations like Hexchar façade panels (70% biochar), partnerships with Audi (storing 10 tons of CO_2 via dealership panels), and HandM (biochar-based sunglasses). Current projects include Sequoia/MOA Board, a fireproof biochar thermoset for construction, and Gen 1 scaling for industrial fillers.

MOA uses high-temperature chars with low volatile content and PAHs, produced via augers, retorts, fluidized beds, and gasifiers. Their materials reach 25% biochar content in polymers and even higher in construction, enabling significant GWP reductions. Even at 20% biochar, many recycled compounds can achieve net-zero emissions, while maintaining compatibility with existing industrial processes.

Supply Chain Challenges and Solutions

According to Daniel Schwaag, co-founder of Made of Air, most biochar still comes from agricultural sources and varies widely in quality, particle size, and feedstock. The sector often lacks understanding of long-term pricing and futures, leading to unstable supply and underestimated distribution costs. Biochar contracts remain short-term and volatile, posing challenges for industrial integration, risk management, and ROI planning.

To address this, Made of Air is working with suppliers to standardize quality and product definitions for industrial use. They are also exploring specialized supply chains and partnerships with wholesalers to manage contracts, streamline logistics, and enable economies of scale through basic processing like grinding and sieving.

Future Applications

The future of biochar depends on converting a variety of biorefinery outputs e.g. energy, biocarbon into high-value, competitive materials that can integrate into mainstream industrial processes. Success will require localized partnerships at biomass sources and the creation of products that can compete with low-cost commodities priced under $200 per ton.

---

[7] https://www.novocarbo.com/

[8] https://wakefieldbiochar.com/?srsltid=AfmBOop2Q-R_TNjtYSFBytDl-KopnbVWLmpiS1ayYC1JIAU2QEC_mE02

[9] https://www.madeofair.com/

Advances in monitoring and verification will help bridge the gap between low-quality offsets and high-cost biochar removals by proving long-term carbon storage in industrial products and infrastructure. Key growth areas include circular material systems, such as reusable packaging, with ongoing work to resolve food-grade certification and PAH concerns.

Ultimately, biochar's widespread adoption will rely on meeting industrial performance standards, not just redefining how carbon removals are measured and monetized.

Innovative Products

MOA pursues the idea of char functionalization. That is, the ability to tune chars' compatibility with any binder system. The company sees this as way to increase fill grades in unfilled materials, while preserving properties of unfilled materials. They are looking to create full utilization of biomass inputs, in terms of char, platform chemicals, and energy, to produce higher value products that can find higher value applications farther away from the biomass source.

4.2 Construction Industry

Biochar has emerged as a transformative material in the construction industry, offering solutions for carbon reduction, waste valorization, and enhanced material performance. Forestry residues (wood chips, sawdust), biosolids, agricultural waste (rice husks, coffee grounds, coconut shell), and nutshells (macadamia, pecan) dominate feedstock choices for use in construction materials. Optimum strength and adsorption properties are achieved at 600–700°C. Modular systems enable decentralized production (Żukiewicz-Sobczak et al. 2020).

Innovative biochar amendments have been developed in concrete additives, insulation materials, fire retardants, and humidity control. Biochars can replace 5–10% of cement, improving compressive strength (up to 24%) and sequestering CO_2. They are mixed with plaster or polymers for thermal regulation and noise reduction. They have been shown to enhance fire resistance in concrete and wood composites. Furthermore, they are also used to absorb moisture to prevent mold in walls and flooring (Mensah et al. 2021; Zhang et al. 2022).

Solid Carbon (US) adds biochar to concrete flooring. Wakefield BioChar has used coffee-ground biochar to replace sand in concrete. Novocarbo produces biochar-based building panels with 80% biochar content for carbon-negative construction. Ecolocked[10] in Germany incorporates biochars in paving stones, precast concrete elements and ready mixed concrete (RMC). Innovative building materials have been developed at several universities, such as the Rochester Institute of Technology[11] and Heriot-Watt University[12].

---

[10] https://www.ecolocked.com/

[11] https://www.rit.edu/

[12] https://www.hw.ac.uk/

Biochars that perform well as building materials have high porosity, good thermal conductivity, high carbon content, and a low hydrogen to organic carbon ratio. A porosity of 59%–65% provides humidity regulation and reduces the weight of composites. Thermal conductivity for insulation is <0.1 W/mK. Carbon content of 70%–90% is preferred for stability and carbon sequestration. Low H/C ratios (0.4) ensure long-term carbon persistence (Tomczyk et al. 2020; Wang et al. 2023; Water et al. 2010).

Biochar-based products must compete with the lower upfront costs of concrete, but they have lower life cycle savings. Fiberglass and polystyrene lack the carbon sequestration benefits of biochar. Biochar-based construction materials can have high production costs. Certifications are emerging through American National Standards Institute (ANSI), American Society for Testing and Materials (ASTM), and European certification. Projects like CHARBUILD[13] are testing the long-term performance of biochar-amended materials in extreme conditions. Companies like ZeroTwelve are developing biochar amended concrete products.[14]

Biochar bridges sustainability and functionality in construction, reducing emissions while enhancing material performance. Projects like Novocarbo and Wakefield BioChar demonstrate scalable solutions, though cost barriers and standardization remain hurdles. With advancements in pyrolysis technology and policy support, biochar could sequester 2.6 billion tons of CO_2e annually by 2050, making it a cornerstone of green construction (IPCC 2022).

5. Part III: Environmental Services

5.1 Decontamination and Remediation

This comprehensive analysis examines real-world applications, market dynamics, and technological innovations driving biochar adoption in environmental remediation.

Soil and Water Remediation Applications

Biochar effectively immobilizes heavy metals (Pb, Cd, Cu) and persistent organic pollutants (DDT, PAHs) through adsorption and pH-dependent precipitation, reducing toxin bioavailability by 30–70% and enabling safe land reuse. (Al-Wabel et al. 2015; Cao et al. 2009; Cao et al. 2011; Uchimiya et al. 2010). The effectiveness of biochar as a contaminant removal play a role in sustainable stormwater management. Integrated into sand filters, bioretention systems, bioswales, and permeable pavements, biochar—often derived from wood waste—has been successfully applied in real-world settings to mitigate urban runoff pollution (Hasan et al. 2011; Koivusalo et al. 2023; Kranner et al. 2019; Pritchard et al. 2022). Performance data show up to 95% heavy metal removal

[13] https://www.dti.dk/projects/charbuild/45696

[14] https://www.zerotwelve.com/

and 80–90% *E. coli* reduction (Matthiesen et al. 2018; USBI 2024; Ding 2026). Notable implementations include modular filters in Oregon and Washington that remove zinc, cadmium and copper.by companies Stormwater Biochar[15] and products derived from high carbon flyash by Oregon Biochar Solutions[16]. Engineered biochars are used for roadside stormwater filtration, and urban soil remediation in the Chesapeake Bay[17].

PFAS and heavy metal removal: Several municipalities in the US remove PFAS/PFOS by carbonizing biosolids. The University of North Carolina at Greensboro (UNCG) has developed nano-engineered biochar for PFAS and heavy metal adsorption from water. The Natural Resources Research Institute (NRRI) has successfully demonstrated PFAS removal from stormwater, while Standard Biochar Corporation has provided biochar for PFAS removal in Maine[18].

Technical Specifications and Production

Physical Properties: Effective remediation biochars exhibit high surface areas (300–600 m^2/g) and optimal porosity, particularly micro and mesopores that enhance contaminant adsorption. Carboxyl and hydroxyl functional groups enable chemical bonding with contaminants like Pb^{2+} and Cr. Alkaline biochars effectively immobilize metals in acidic soils, reducing leaching, while low H/C ratios (<0.4) ensure long-term carbon persistence (Bian et al. 2022; Lobzenko et al. 2022; Manna et al. 2015; Nguyen et al. 2024).

Alkaline biochar can be achieved by producing it at high temperature (600–800°C) from pyrolysis, gasification, or combustion of wood, eggplant waste, Manure, poultry litter and nutshells (He et al. 2024; Murtza et al. 2024). Gasification produces energy-rich syngas alongside biochar, suitable for decentralized systems. Chemical activation (KOH, H_3PO_4) or steam activation increases surface area exceeding 50% (Lua et al. 2004; Ahmedna et al. 2000). Biochars combined with biological agents including earthworms, dung beetles, and soil microorganisms enhance soil reclamation effectiveness.

Industry Innovation: Bioforcetech Corporation Case Study

BioForcetech[19] Corporation, established in 2012 and headquartered in South San Francisco, represents a leading innovator in sustainable environmental solutions. The company has developed a proprietary dual process combining BioDrying and Pyrolysis of biosolds from waste treatment without additives or bulking agents like biomass. They have installations in California and Pennsylvania.

---

15 https://stormwaterbiochar.com/
16 https://www.chardirect.com/
17 https://www.scalingupbiochar.com
18 https://www.standardbiocarbon.com/news-blog/tackling-pfas-contamination-of-soil-in-unity-maine-using-biochar
19 https://bioforcetech.com/

The BioDrying system treats biosolids (typically 20% solids, 80% water) using naturally occurring mesophilic or thermophilic bacteria under optimized conditions. This biological process achieves temperatures of 60–65°C, increasing total solids from 20% to 80–90% within 2.5 days while providing 70% energy savings compared to conventional drying. The aerobic process prevents H_2S production, with ammonia managed through scrubber and biofilter systems.

Dried biosolids undergo pyrolysis at 650°C for 15–20 minutes, producing high-quality biochar while eliminating microplastics, per- and polyfluoroalkyl substances (PFAS) and contaminants of emerging concern (CEC). The resulting dry biochar contains 28–50% carbon and serves as sustainable pigment, replacing carbon black as an ingredient.

Market Position and Scale: BioForcetech operates with unique market positioning, treating all sludge types with capacity for 30,000 wet tons annually, targeting municipalities processing 1–10 million gallons average daily flow. The company is scaling across Italy and the US, addressing the global biosolids management challenge where US biosolids are currently managed through landfills, land application, and 8% incineration.

Remediation

Market Outlook and Environmental Impact: Biochar represents a convergence of environmental remediation and circular economy principles, offering scalable solutions for soil and water decontamination. Successful projects including Chalmers University's DDT remediation and PYREG's[20] carbon-negative systems demonstrate biochar's potential to simultaneously mitigate pollution and sequester carbon. Policy support and technological innovation are addressing cost and scalability challenges, positioning biochar as a cornerstone technology for sustainable environmental management.

The remediation market shows significant growth potential as regulatory pressures increase, and climate change mitigation strategies prioritize carbon-negative technologies. Biochar's versatility across multiple contamination scenarios, combined with its economic and environmental co-benefits, establishes it as a critical tool for environmental restoration and sustainable waste management. Biochars and engineered biochar-based media can replace activated carbons. They can be used effectively in geotechnical applications to improve soil hydrology. Market expansion requires demonstration and adoption of specifications for biochar that contractors can buy and install in remediation sites stormwater facilities.

Future Development: BioForcetech has initiated collaboration with USDA to create a comprehensive database of biochar characteristics corresponding to specific feedstocks, enabling targeted market applications and standardization.

[20] https://pyreg.com/

6. Part IV: Emerging Applications

6.1 Renewable Energy

Biochar has emerged as a transformative force in renewable energy systems, offering integrated solutions for carbon sequestration, waste valorization, and energy efficiency. This analysis examines the technology's applications, economic impacts, and market dynamics across the renewable energy sector through comprehensive research and industry interviews.

Production Technologies and Energy Integration

Co-production from Energy Facilities: Biochar serves as a valuable co-product in large biomass energy facilities, where high-carbon chars are recovered from gasifiers used for direct heat generation in wood product drying and from boilers generating process steam. While char yield from these processes can range from 2–8% of fuel input for staged combustion to 15% for gasification, this approach captures carbon that would otherwise reduced to ash, creating new products and additional environmental benefits that can add revenue to plants that operate off of very narrow profit margins.

Primary Production Methods: Biochar production utilizes pyrolysis, gasification or combustion of diverse feedstocks including agricultural residues (rice husks, corn stalks), forestry waste (wood chips, sawdust), nutshells, and sewage sludge. Slow pyrolysis and gasification have emerged as the predominant technologies for optimizing biochar yield, typically achieving 15–25% of biomass input conversion. The process simultaneously recovers condensable tars, oils, and liquids alongside non-condensable gases, with liquids typically combusted with gases for energy generation. Production systems employ various reactor configurations including moving grates, fixed beds, and fluidized beds.

Fast Pyrolysis Applications: Fast pyrolysis primarily targets oil recovery for refining into diesel or sustainable aviation fuels, producing biochar as a secondary product (15% yield). While this biochar is often combusted to drive the pyrolysis process, recovery for alternative applications is increasingly implemented where economically viable.

Leading Industry Applications

Integrated Energy-Biochar Systems: Syncraft (Austria)[21], Oregon Biochar Solutions, Pacific Biochar[22], and PYREG (Germany) exemplify successful biomass-to-biochar conversion while generating syngas for heat and power applications, achieving carbon offset potential of 2.8 mt CO_2 per metric ton of biochar produced. Wakefield BioChar has demonstrated innovative feedstock

[21] https://www.syncraft.at/en/
[22] https://pacificbiochar.com/

utilization by converting coffee grounds into biochar and syngas for decentralized energy systems.

Advanced Hydrogen Production: Haffner Energy's Hynoca® technology[23] represents cutting-edge innovation in biomass thermolysis, producing renewable hydrogen alongside biochar while sequestering 12 kg CO_2 per kg of H_2 generated. This dual-benefit approach addresses both clean fuel production and carbon sequestration objectives.

Energy Storage Applications: Biochar can be produced with high surface area (>300 m^2/g) and electrical conductivity which enable applications in energy storage systems, serving as graphite replacement in biochar-based supercapacitors and lithium-ion batteries, contributing to renewable energy storage infrastructure development.

Additionally, synthesis gases generated from pyrolysis processes can provide direct fossil fuel replacements, while biochar-amended insulation materials reduce Heating, Ventilation, and Air Conditioning (HVAC) energy consumption by 20–30%, demonstrating multiple pathways for energy efficiency enhancement.

Technical Specifications and Quality Parameters

High-Carbon Boiler Char Characteristics: Biochars recovered from biomass gasification and combustion exhibit optimal properties including high carbon content (70–90%), low H/C ratios (<0.4) ensuring stability and calorific value, substantial porosity (58–65%) enhancing adsorption capacity for energy storage applications, and minimal ash content (Lehmann and Joseph 2015; Brewer et al. 2014).

Process Optimization: Slow pyrolysis at 500–700°C with off-gas combustion maximizes biochar yield (25–35%) and energy efficiency. Gasification processes operating above 750°C produce syngas with up to 15% biochar yield, while combustion achieves low carbon yield (1–2%) but produces high-quality biochar characteristics (Antal and Grønli 2003; Mohan et al. 2014).

Modular System Development: Multiple companies have developed modular, decentralized systems typically operating at 24–48 tons per day capacity, enabling distributed production and reducing transportation costs while maintaining economies of scale (Woolf et al. 2010; Scholz et al. 2014).

Industry Case Study: SYNCRAFT – Reverse Power Plant Innovation

Company Profile and Technology: SYNCRAFT, a leading Austrian green technology company based in Tyrol, has pioneered "reverse power plants" that generate affordable green energy while achieving negative CO_2 emissions. Co-founder and Managing Director Marcel Huber emphasize the company's focus

---

[23] https://www.haffner-energy.com/haffner-energy-unveils-hynoca-flex-500-ig-a-flexible-cost-effective-alternative-to-grey-hydrogen-and-fossil-fuels/

on converting valuable forest residues into heat, electricity, and green carbon through resource-efficient processes.

Production Metrics and Distribution: SYNCRAFT's reverse power plants demonstrate impressive conversion efficiency, transforming 100% fuel input (dry basis) into 27–28% electricity, 40% heat, and 25% carbon. For example, one ton of fuel wood chips with five megawatts energy content (dry basis) produces 2 megawatts of heat equivalent and roughly 180 kilograms of pure carbon with up to 90% fixed carbon.

Scale and Market Reach: By 2024, SYNCRAFT realized over 40 climate-positive reverse power plants for industry and communities worldwide, including installations in Japan, Italy, and Croatia. Combined production from all plants reached 30,000 tons of biochar annually, with customers primarily utilizing biochar as soil amendment in non-agricultural sectors, particularly EPC (Engineering, Procurement, and Construction) projects, with construction materials representing the second-largest application stream.

Product Specifications and Safety: The biochar produced for energy generation is sized below 4 mm, with critical safety protocols implemented to prevent auto-ignition during transportation. The company moistens biochar to approximately 20% water content using mixing systems at the production process conclusion, as failure to implement this step creates significant fire risk during transportation in big bags.

Project Development Philosophy: SYNCRAFT's project development centers on client electricity and heat demand, with Marcel Huber noting, "we don't build a reverse power plant when you don't need the power". Projects targeting breweries exemplify this approach, where baseload heat demand enables very cheap heat supply with primary revenue from electricity generation. The company no longer pursues projects without heat utilization, recognizing that combined heat, power, biochar, and carbon credit sales create more resilient project economics.

Market Drivers: High electricity prices serve as the primary driver for biochar plant construction necessity, with heat and biochar representing secondary benefits. However, projects incorporating heat, power, biochar, and carbon credit sales demonstrate significantly enhanced resilience and economic viability.

Industry Case Study: Oregon Biochar Solutions – Regulatory and Market Challenges

Company Mission and Approach: Oregon Biochar Solutions (OBS)[15], led by Karl Strahl, represents a pioneering approach to sustainable biochar production and utilization. The company addresses environmental challenges by recovering high quality biochar from converting biomass to power through advanced combustion processes, while generating renewable energy. The biochars are upgraded and used in a variety of applications, simultaneously capturing and sequestering carbon and enhancing soil health. High carbon boiler char from biomass plants are major contributors to the voluntary carbon markets, providing an important service of climate mitigation which helps finance carbon recovery and sequestration.

Regulatory Framework Challenges: Strahl identifies significant regulatory clarity gaps regarding biomass classification as renewable energy, without considering key regulatory framework details. His biochar plant produces 6,000 tons of biochar

While biomass power plants reduce both anthropogenic and biogenic emission and hazardous air pollutants and particulate, stringent regulatory requirements for reduction of criteria pollutants such as nitrogen oxides (NOx) without recognizing broader environmental benefits may cause biomass facilities to close.

Environmental Impact Quantification: Strahl advocates for regulatory focus on comprehensive environmental impact assessment, noting that plant shutdowns could increase inefficient waste burning while his plants currently provide electricity to approximately 26,000 homes. The challenge involves quantifying biochar production value while managing emission controls, electricity generation, and operational integration requirements.

Feedstock Optimization: While biomass facilities can accommodate a wide range of moisture contents OBS utilizes mostly dry materials with approximately 30% moisture content or less, as significant energy is required for moisture removal when heating to temperatures as high as 950°C. Optimal feedstocks for biochar production in renewable energy applications include wood residues and agricultural waste, particularly nut shells, which provide efficient conversion rates and high-quality biochar output.

Market Challenges and Future Perspectives

Market Maturity Requirements: Industry advancement requires establishing mature markets, with current US producers often lacking effective marketing strategies and appropriate product packaging and distribution. Recognition that biochars serve as raw materials or ingredients rather than final product necessitates market development for finished product transformation and distribution.

Economic Drivers and Barriers: Global demand exists for fuel, heat, and power without country-specific limitations, but economic factors present significant challenges when fossil fuels maintain low prices, reducing alternative energy transition incentives. Beyond power and heat generation, biochar production offers sustainable replacement gas potential for high-temperature industrial processes requiring high temperatures, particularly attractive to cement and other large-scale operations seeking lower-cost fuels and reduced carbon footprints.

Industrial Applications and Opportunities: Pyrolysis gas can support cement production decalcification steps and provide necessary carbon for achieving final rotary kiln temperatures, significantly reducing CO_2 emissions. The steel industry represents a strong ally due to direct sustainable carbon source requirements, with biochar serving as ideal green carbon for metallurgical applications. Some large biocarbon facilities, like the 160,000 tons per year Steel Dynamics Inc./SDI Carbon Solutions LLC facility in Mississippi may also produce biochars[24].

Value Chain Development: Progress depends on proper energy valorization, as energy undervaluation hinders bio-based solution adoption. Higher-value markets beyond agriculture, including urban soil amendments and biochar-enriched concrete products, present significant opportunities for expanding biochar utilization and improving project economics.

Technology Solutions for Implementation Challenges

High capital and production costs in the US have been partly addressed through USDA grants and modular pyrolysis systems. Feedstock variability challenges are managed through process control to produce products that meet product requirements, standards, and certificates (EBC, C-Sink) ensuring quality consistency. Durability improvements utilize engineered composites such as MgO-biochar enhancing stability in harsh operational conditions. Mobile and modular pyrolysis systems promise to enable local production, reducing logistics barriers and transportation costs.

Strategic Market Position and Future Outlook

Biochar effectively bridges renewable energy and circular economy objectives, delivering carbon-negative solutions from waste biomass streams. Successful projects including PYREG and Hynoca® demonstrate scalable models for hydrogen production and energy storage applications, while policy support through carbon credits accelerates market adoption.

Despite ongoing challenges in markets, cost management and standardization, technological advancements in pyrolysis systems position biochar as a cornerstone technology in the global energy transition. The integration of biochar production with renewable energy generation creates multiple revenue streams while addressing critical environmental challenges, establishing a foundation for sustainable energy infrastructure development and carbon sequestration at scale.

The industry's evolution toward mature markets, combined with supportive regulatory frameworks and technological innovation, can position biochar as an essential component of renewable energy systems, offering economically viable solutions for waste valorization, carbon sequestration, and clean energy production.

6.2 Pharmacy Potential

The potential of biochar in the pharmaceutical sector is related to its high surface area and porosity to remove pollutants like antibiotics and hormones

[24] https://ir.steeldynamics.com/steel-dynamics-announces-location-of-planned-biocarbon-production-operations-a-meaningful-strategic-ghg-reduction-initiative/

from wastewater and to support controlled, long-term drug release (Chauhan et al. 2023). Biochar has the potential to aid pharmaceutical manufacturing by adsorbing toxins (Zhang et al. 2024). The adsorption efficiency is remarkably high, with "recent studies demonstrate that biochar can achieve >90% removal efficiency for heavy metals such as lead and cadmium, >85% adsorption capacity for organic pollutants such as dyes and phenols" (Kumar et al. 2024), biochar also helps combat antimicrobial resistance (Johnson et al. 2023). However, comprehensive studies specifically demonstrating all these applications integrated into pharmaceutical manufacturing processes appear to be emerging research areas that may need further investigation and documentation.

Dr. Reji Kurien Thomas, CEO of TOL Biotech[25] leads studies of pharmaceutical-grade biochar made from medicinal plant waste, they have been studying the potential of biochar for pharmaceutical applications by integrating biotechnology, AI, and quantum science to create applications in drug delivery, detoxification, cancer treatment, and regenerative medicine.

They consider that clean biomass with low toxin and heavy metal content is essential, ideally rich in lignin, cellulose, and hemicellulose for stable carbon structures. For that the highest purity and quality are fundamental. Therefore, it is important to eliminate or remove any trace contaminants or residues, discoloration, or odor that may be present (Chauhan et al. 2023).

Pharmaceutical-grade biochar is primarily produced through pyrolysis at 400–500°C for 2–4 hours in an inert atmosphere. Steam activation occurs at 800–900°C with controlled steam injection. Cooling and washing remove residual ash and impurities. Acid treatment involves HCl washing to remove metal impurities. Final purification includes multiple water washes until the pH is neutral. Pharmaceutical grade activated carbon is highly purified to meet pharmacopeia standards, such as the United States Pharmacopeia (USP) and European Pharmacopoeia (EP). It is available in various micron sizes to optimize adsorption efficiency in different drug formulations, and it maintains chemical stability (Angın et al. 2013).

Biochar-based pharmaceutical products have sizable long-term market potential, especially for drug delivery and advanced biomaterials. However, this potential has not yet been commercially realized; the segment is at the research/innovation stage and may take several years before impacting the broader global biochar or pharmaceutical markets in a measurable way. Recognized market opportunities exist—but are fundamentally tied to future research and development breakthroughs and regulatory progress.

Animal Feed and Stable Litter

Biochar is an emerging material in livestock management, used both as a feed additive and stable amendment. When added to animal feed at 0.3–1% of dry

[25] https://www.tol-biotech.com/pages/medical-healthcare-industries

matter, it has been shown to improve digestion, bind toxins, and reduce methane emissions by 10–18% (Leng et al. 2012; Li et al. 2018). As a bedding additive (5–10% by volume), biochar reduces ammonia emissions by up to 80%, helps control moisture, and lowers the incidence of hoof diseases and respiratory issues (Petersen et al. 2017; Wang et al. 2020). These benefits enhance animal health and productivity, increasing meat, egg, and milk yields (Steiner et al. 2010; Zhang et al. 2021). In developing countries organizations like Warm Heart Worldwide, Biochar Life and others have shown thousands of smallholders the benefits of biochar. Biochars apparently improve animal health by reducing disease, leading to improved production of eggs and milk.

Biochar made from agricultural and forestry residues—such as rice husks, nut shells, wood chips, and poultry litter—is preferred for these applications due to its high porosity (>50%), surface area (>300 m^2/g), alkaline pH (8–10), and low H/C ratio, which ensures durability and toxin-binding capacity (Lehmann and Joseph 2015; Mukherjee and Zimmerman 2013). Low-temperature pyrolysis (around 450°C) and top-lit updraft gasifiers are commonly used production methods that optimize these properties (Bridgwater 2012; Tan et al. 2015).

Economically, biochar lowers veterinary costs, reduces the need for bedding replacement, and provides access to carbon credits, making it a cost-effective alternative to activated carbon. Challenges include regulatory uncertainty, quality variation, cost, and limited consumer familiarity. However, successful cases like those from CharLine show that with improved pyrolysis technologies and policy support, biochar can reduce livestock sector emissions while boosting farm profitability.

Product development using biochar has proven to be an excellent way to share its benefits with the population, make biochar production economically viable, and generate jobs.

CharLine Case Study

Dominik Dunst and Gerald Dunst from CharLine[26] were interviewed and they have exemplified this by developing a comprehensive line of biochar products specifically for animal feed.

The animals benefiting from these products include dogs, cats, horses, cattle, calves, pigs, poultry, and sheep. These products primarily enhance the animals' gastrointestinal health and metabolism, with notable benefits such as a reduction in diarrhea. Achieving these results requires rigorous testing, ensuring the final biochar meets certification standards such as the European Biochar Certification (EBC) for animal feed[27]. At CharLine, meticulous care is taken in selecting feedstocks, whether a mix of different types or a single feedstock, depending on the product's intended use. The biochar is pyrolyzed at high temperatures, often between 700°C and 900°C, depending on the feedstock and the goal you want

---

[26] https://www.char-line.com/

[27] https://www.european-biochar.org/en/ct/2-EBC-and-WBC-guidelines-documents

to achieve. According to Dominik, the key to producing high-quality biochar for animal feed lies in selecting the right feedstock, controlling the residence time in the reactor, and maintaining optimal pyrolysis temperatures. Additionally, the form of application is crucial. For instance, larger particle sizes are recommended for cattle feed.

Sonnenerde of GmbH, the parent company of CharLine, emphasizes the significance of the biochar-mineral complex, a perspective shaped by their collaboration with Dr. Stephen Joseph[28]. Practical tests conducted by the company have demonstrated that using the biochar-mineral complex requires only half the amount of biochar to achieve equivalent results in odor reduction in liquid manure. Additionally, this approach appears to enhance the efficacy of feed char, requiring smaller quantities of biochar to achieve similar outcomes.

Typically, CharLine (FeedChars and ManureChars) are CharLine products. It is recommended a 0.5% biochar application rate by weight, either as an amendment to animal feed or slurry. For these use cases, the nutrient content of the biochar is less critical. Instead, the fineness of the biochar plays a pivotal role—finer biochar delivers better results. However, the production technique must preserve the porosity of the biochar, even for particles smaller than 1 mm, to ensure optimal performance.

The use of biochar in animal feed also offers economic advantages to producers. For example, in Austria, a turkey farmer typically incurs veterinary and medication costs of at least €10,000 per visit. By integrating biochar into animal feed, these expenses can be significantly reduced, "by the use of biochar feed it's not necessary to use antibiotics anymore, but the performance in the animal husbandry production is still the same", additionally the animal health increase, and the veterinary costs fall at the same time. It is very economical to use feed char, but there are still very few farmers who use them. "By adding biochar to the slurry or as bedding material in the animal houses, less nitrogen fertilizer is needed", said Gerald Dunst.

Energy costs remain a notable challenge for biochar production in Austria. Nevertheless, the market remains profitable, with prices in Austria averaging $872 per ton for general biochar and $1,526 per ton for feed-grade biochar. Additionally, carbon credits offer an important supplementary revenue stream for producers.

Regarding competitive products there is no material with very similar properties, except for Leonardite which is a naturally occurring, soft, brown or black mineral-like substance that is the result of the oxidation of lignite (brown coal). It is a fossil, which can be used also as feed amendment, also some stone powders are used for similar applications.

Among the various markets, the poultry sector shows significant promise for biochar-based animal feed products. Biochar feed products for aquaculture are still in the early stages of development, with a potential market beginning to take shape. However, the high costs associated with the fish industry may present challenges for conducting the necessary testing to establish these products.

[28] https://scholar.google.com/citations?user=6m67sQUAAAAJ&hl=en

7. Critical Analysis and Conclusion

Biochars show strong potential across multiple sectors, but key barriers continue to limit their large-scale adoption. The most common challenge is high production cost, which affects agriculture, industry, animal feed, and environmental remediation alike. These costs stem from expensive pyrolysis systems, inconsistent feedstocks, and limited economies of scale. In industry, challenges also include fluctuating quality due to lack of post processing leading to issues like particle size variability, unstable pricing, short-term contracts, and underestimated logistics and distribution costs.

Standardization is another cross-cutting issue. Without clear, consistent specifications, it's difficult to guarantee biochar quality for specific applications, especially in pharmaceuticals and construction. Regulatory uncertainty further delays market growth. A lack of public awareness and technical knowledge, particularly among farmers and policymakers, slows the development of supportive policies and broader adoption. The European Biochar Certificates (EBC) lead the specifications for biochar use. Educational initiatives like CBEN, Biochar Europe, Australia New Zealand Industry Group (ANZBIG), US Biochar Initiative (USBI), US Biochar Coalition (USBC), International Biochar Initiative (IBI) and other groups are helping, but wider outreach is needed.

Despite these barriers, sector-specific innovations are emerging. In agriculture, biochar reduces chemical fertilizer use and water demand, but nutrient enrichment is sometimes necessary before it is applied. In industry, supply chain improvements and contract structures can stabilize pricing and quality. In environmental remediation, engineered biochars like MgO-biochar improve durability and performance. In energy, biochar's integration with renewable systems supports carbon sequestration and creates new revenue streams. In pharmaceuticals, smart drug delivery systems, 3D-printed biochar scaffolds, and blockchain-based traceability show high potential, though scale-up and regulatory approval remain challenges for market entry. For animal feed, poultry shows promise, but aquaculture faces cost barriers for testing and product development.

In conclusion, biochar is a powerful tool for sustainable development, but adoption depends on reducing production costs, creating clear standards, improving public knowledge, and securing fair carbon credit pricing. Coordinated action across sectors along with investment in innovation, education, and policy will be key to unlocking its full impact.

Acknowledgment

The authors would like to express their gratitude to all interviewees for generously sharing their processes, observations, and knowledge. Your insights and experiences have been invaluable, bringing depth and richness to this chapter. This work would not have been possible without your contributions.

References

Ahmedna, M., and Johns, M.M. 2000. Preparation of activated carbons from groundnutshell by chemical activation with KOH and K_2CO_3. *Carbon* 38(3): 595–601. https://doi.org/10.1016/S0008-6223(99)00156-X

Al-Wabel, M., Usman, A., El-Naggar, A., Aly, A., Ibrahim, H., El-Maghraby, S. et al. 2015. Conocarpus biochar as a soil amendment for reducing heavy metal availability and uptake by maize plants. *Saudi Journal of Biological Sciences* 22(4): 503–511. https://doi.org/10.1016/j.sjbs.2014.12.003

Angın, D., Altintig, E., and Köse, T.E. 2013. Influence of process parameters on the surface and chemical properties of activated carbon obtained from biochar by chemical activation. *Bioresource Technology* 148: 542–549. doi: 10.1016/j.biortech.2013.08.164

Antal, M.J., and Grønli, M. 2003. The art, science, and technology of charcoal production. *Industrial and Engineering Chemistry Research* 42(8): 1619–1640. https://doi.org/10.1021/ie0207919

Beilei, W., Yunchang, P., Longxin, L., Dongliang, Z., Li, M., Ligeng, J. et al. 2023. Drivers of biochar-mediated improvement of soil water retention capacity based on soil texture: A meta-analysis. *Geoderma* 437: 116591. https://doi.org/10.1016/j.geoderma.2023.116591

Bian, P., Liu, Y., Zheng, X., and Shen, W. 2022. Removal and mechanism of cadmium, lead and copper in water by functional modification of silkworm excrement biochar. *Polymers* 14(14): 2889. https://doi.org/10.3390/polym14142889

Brewer, C.E., Chuang, V.J., Masiello, C.A., Gonnermann, H., and Gao, X. 2014. New approaches to measuring biochar density and porosity. *Biomass and Bioenergy* 66: 176–185. https://doi.org/10.1016/j.biombioe.2014.03.020

Bridgwater, A.V. 2012. Review of fast pyrolysis of biomass and product upgrading. *Biomass and Bioenergy* 38: 68–94. https://doi.org/10.1016/j.biombioe.2011.01.048

Bruun, E., Petersen, C., Hansen, E., Holm, J., and Hauggaard-Nielsen, H. 2014. Biochar amendment to coarse sandy subsoil improves root growth and increases water retention. *Soil Use and Management* 30(1): 109–118. https://doi.org/10.1111/sum.12102

Cao, X., Lena, Q., Gao, B., and Harris, W. 2009. Dairy-manure derived biochar effectively sorbs lead and atrazine. *Environmental Science and Technology* 43(9): 3285–3291. https://doi.org/10.1021/es803092k

Cao, X., Lena, Q., Yuan, L., Gao, B., and Harris, W. 2011. Simultaneous immobilization of lead and atrazine in contaminated soils using dairy-manure biochar. *Environmental Science and Technology* 45(11): 4884–4889. https://doi.org/10.1021/es103752u

Chauhan, S., Shafi, T., Dubey, B.K., and Chowdhury, S. 2023. Biochar-mediated removal of pharmaceutical compounds from aqueous matrices via adsorption. *Waste Disposal & Sustainable Energy* 5(1): 37–62. doi: 10.1007/s42768-022-00118-y.

Crespo-Barreiro, A., Gómez, N., González-Arias, J., Ortiz-Liébana, N., González-Andrés, F., and Cara-Jiménez, J. 2023. Scaling-up of the production of biochar from olive tree pruning for agricultural use: Evaluation of biochar characteristics and phytotoxicity. *Agriculture* 13(5): 1064. https://doi.org/10.3390/agriculture13051064

de Jesus Duarte, S., Glaser, B., and Cerri, E.E.P. 2019. Effect of biochar particle size on physical, hydrological, and chemical properties of loamy and sandy soils. *Agronomy* 4(9).

Ding, Y., Liu, Y., Liu, S. et al. 2016. Biochar to improve soil fertility. A review. *Agronomy for Sustainable Development* 36: 36. https://doi.org/10.1007/s13593-016-0372-z

Earth.org. 2024. *The Remarkable Growth of the Global Biochar Market: A Beacon of Environmental Progress*. Available at: https://earth.org/the-remarkable-growth-of-the-global-biochar-market/

Hasan, M., Vasquez, R., and Geza, M. 2021. Application of biochar in stormwater treatment: Experimental and modeling investigation. *Processes* 9(5): 860. https://doi.org/10.3390/pr9050860

He, D., Luo, Y., and Zhu, B. 2024. Feedstock and pyrolysis temperature influence biochar properties and its interactions with soil substances: Insights from a DFT calculation. *Science of the Total Environment* 922: Article 171259. https://doi.org/10.1016/j.scitotenv.2024.171259

He, Y., Zhou, X., Jiang, L., Li, M., Du, Z., Zhou, G. et al. 2017. Effects of biochar application on soil greenhouse gas fluxes: A meta-analysis. *GCB Bioenergy* 9(4): 743–755.

Huang, Y., Li, B., Liu, D., Xie, X., Zhang, H., Sun, H. et al. 2020. Fundamental advances in biomass autothermal/oxidative pyrolysis: A review. *ACS Sustainable Chemistry and Engineering* 8(32): 11888–11905. https://doi.org/10.1021/acssuschemeng.0c04196

Ibitoye, S., Loha, C., Mahamood, R., Jen, T., Alam, M., Sarkar, I. et al. 2024. An overview of biochar production techniques and application in iron and steel industries. *Bioresources and Bioprocessing* 11(1). https://doi.org/10.1186/s40643-024-00779-z

IPCC. 2022. *Climate Change: Mitigation of Climate Change*. Contribution of Working Group III to the Sixth Assessment Report of the Intergovernmental Panel on Climate Change. Cambridge University Press. https://doi.org/10.1017/9781009157926

Jeffery, S., Verheijen, F.G., van der Velde, M., and Bastos, A.C. 2011. A quantitative review of the effects of biochar application to soils on crop productivity using meta-analysis. *Agriculture, Ecosystems and Environment* 144(1): 175–187.

Jia, X., Wang, Y., Li, L., Tan, W., Wang, S., Rajendran, M. et al. 2021. A critical review of biochar-based nitrogen fertilizers and their effects on crop production and the environment. *Reviews in Environmental Science and Bio/Technology* 20(3): 677–698. https://doi.org/10.1007/s11157-021-09584-4

Johnson, K.L., Smith, D.R., and Brown, A.M. 2023. Biochar-mediated mitigation of antimicrobial resistance in environmental systems. *Water Research* 231, 119645.

Koivusalo, H., Dubovik, M., Wendling, L., Assmuth, E., Sillanpää, N., and Kokkonen, T. 2023. Performance of sand and mixed sand–biochar filters for treatment of road runoff quantity and quality. *Water* 15(8): 1631. https://doi.org/10.3390/w15081631

Kranner, B., Afrooz, A., Fitzgerald, N., and Boehm, A. 2019. Fecal indicator bacteria and virus removal in stormwater biofilters: Effects of biochar, media saturation, and field conditioning. *PLOS One* 14(9): e0222719. https://doi.org/10.1371/journal.pone.0222719

Kumar, S., Patel, R., and Gupta, M. 2024. Advanced biochar materials for environmental remediation and pharmaceutical applications. *Chemical Engineering Journal* 478: 147234.

Lehmann, J., and Joseph, S. (Eds.). 2015. *Biochar for Environmental Management: Science, Technology and Implementation* (2nd ed.). Routledge.

Leng, R.A., Preston, T.R., and Inthapanya, S. 2012. Use of biochar to improve animal nutrition and reduce enteric methane emissions in ruminants: Review. *Livestock Research for Rural Development* 24(11). http://www.lrrd.org/lrrd24/11/leng24242.htm

Li, X., Mao, S., Jiang, L., Yang, X., Lu, H., Liu, S., and Yin, S. 2018. Effects of dietary biochar on growth performance, nutrient digestibility and methane emission in beef cattle. *Animal Feed Science and Technology* 236: 56–62. https://doi.org/10.1016/j.anifeedsci.2017.11.020

Liu, Z., Dugan, B., Masiello, C., and Torgersen, H. 2017. Biochar particle size, shape, and porosity act together to influence soil water properties. *PLOS One* 12(6): e0179079. https://doi.org/10.1371/journal.pone.0179079

Lobzenko, I., Burachevskaya, M., Zamulina, I., Barakhov, A., Bauer, T., Mandzhieva, S. et al. 2022. Development of a unique technology for the pyrolysis of rice husk biochar for promising heavy metal remediation. *Agriculture* 12(10): 1689. https://doi.org/10.3390/agriculture12101689

Lua, A.C., Yang, T., and Guo, J. 2004. Effects of pyrolysis conditions on the properties of activated carbon prepared from pistachio-nut shells. *Journal of Analytical and Applied Pyrolysis* 72(2): 279–287. https://doi.org/10.1016/j.jaap.2004.02.002

Lyu, H., Zhang, H., Chu, M., Zhang, C., Tang, J., Chang, S. et al. 2022. Biochar affects greenhouse gas emissions in various environments: A critical review. *Land Degradation and Development* 33(17): 3327–3342. https://doi.org/10.1002/ldr.4405

Manna, S., and Singh, N. 2015. Effect of wheat and rice straw biochars on pyrazosulfuron-ethyl sorption and persistence in a sandy loam soil. *Journal of Environmental Science and Health Part B* 50(7): 463–472. https://doi.org/10.1080/03601234.2015.1018757

Matthiesen, E., Nalven, S.G., Spector, D., and Zhang, L. 2018. Innovative filter design application targeting *E. coli* and phosphorus removal. *Forester Network*. https://www.stormwater.com/home/contact/13035619/ed-matthiesen-sarah-nalven-diane-spector-lu-zhang

Mensah, R., Shanmugam, V., Narayanan, S., Razavi, N., Ulfberg, A., Blanksvärd, T. et al. 2021. Biochar-added cementitious materials—A review on mechanical, thermal, and environmental properties. *Sustainability* 13(16): 9336. https://doi.org/10.3390/su13169336

Minasny, B., Salami, S., McBratney, A., and Young, I. 2021. Greater, but not necessarily better: The influence of biochar on soil hydraulic properties. *European Journal of Soil Science* 72(5): 2033–2048. https://doi.org/10.1111/ejss.13105

Mohan, D., Sarswat, A., Ok, Y.S., and Pittman, C.U. Jr. 2014. Organic and inorganic contaminants removal from water with biochar, a renewable, low cost and sustainable adsorbent—A critical review. *Bioresource Technology* 160: 191–202. https://doi.org/10.1016/j.biortech.2014.01.120

Mukherjee, A., and Zimmerman, A.R. 2013. Organic carbon and nutrient release from a range of biochars produced from different feedstocks and pyrolysis temperatures. *Environmental Science and Technology* 47(7): 3571–3578. https://doi.org/10.1021/es303532d

Murtaza, G., Usman, M., Iqbal, J. et al. 2024. Liming potential and characteristics of biochar produced from woody and non-woody biomass at different pyrolysis temperatures. *Scientific Reports* 14: 11469. https://doi.org/10.1038/s41598-024-61974-8

Nguyen, H., Le, P., Phương, N., Hai, T., Nguyen, T., Dinh, T. et al. 2024. Removing emerging organic pollutants (oxytetracycline) by engineered biochars originated from agricultural wastes. *Environmental Progress and Sustainable Energy* 44(1). https://doi.org/10.1002/ep.14512

Petersen, S.O., Sommer, S.G., Amon, B., and Møller, H.B. 2017. Reducing ammonia emissions from animal production—A review. *Animal* 11(5): 776–787. https://doi.org/10.1017/S1751731116002202

Pritchard, J., Hawkins, K., Cho, Y., Spahr, S., Struck, S., Higgins, C. et al. 2022. Black carbon-amended engineered media filters for improved treatment of stormwater runoff. *ACS Environmental Au* 3(1): 34–46. https://doi.org/10.1021/acsenvironau.2c00037

Roberts, K., Gloy, B., Joseph, S., Scott, N., and Lehmann, J. 2009. Life cycle assessment of biochar systems: Estimating the energetic, economic, and climate change potential. *Environmental Science and Technology* 44(2): 827–833. https://doi.org/10.1021/es902266r

Safarian, S. 2023. Climate impact comparison of biomass combustion and pyrolysis with different applications for biochar based on LCA. *Energies* 16(14): 5541. https://doi.org/10.3390/en16145541

Scholz, S.M., Sembres, T., Roberts, K., Whitman, T., Wilson, K., and Lehmann, J. 2014. Biochar systems for smallholders in developing countries: Leveraging current knowledge and exploring future potential for climate-smart agriculture. *World Bank Publications*, No. 88888, Washington DC.

Stavrakou, V., Nikodemou, O., Makri, I., Savvakis, N., Sourkouni, G., and Argirusis, C. 2022. Atmospheric carbon removal via industrial biochar systems: A techno-economic-environmental study. *Journal of Cleaner Production* 364: 132383. https://doi.org/10.1016/j.jclepro.2022.132383

Steiner, C., Teixeira, W.G., Lehmann, J., Nehls, T., Macêdo, J.L.V., Blum, W.E.H. et al. 2010. Microbial exchange in the rhizosphere of biochar-amended soils. *Plant and Soil* 333(1–2): 54–66. https://doi.org/10.1007/s11104-010-0343-5

Tan, K.H., Lee, J.W., and Chew, T.W. 2015. Influence of pyrolysis temperature on properties of biochar from oil palm empty fruit bunch. *Bioresource Technology* 198: 114–121. https://doi.org/10.1016/j.biortech.2015.08.066

Tomczyk, A., Sokołowska, Z., and Boguta, P. 2020. Biochar physicochemical properties: Pyrolysis temperature and feedstock kind effects. *Reviews in Environmental Science and Bio/Technology* 19(1): 191–215. https://doi.org/10.1007/s11157-020-09523-3

Uchimiya, M., Lima, I., Klasson, K., Chang, S., Wartelle, L., and Rodgers, J. 2010. Immobilization of heavy metal ions (CuII, CdII, NiII, and PbII) by broiler litter-derived biochars in water and soil. *Journal of Agricultural and Food Chemistry* 58(9): 5538–5544. https://doi.org/10.1002/etc.4542

US Biochar Initiative. 2024. *Biochar for Stormwater Treatment: Technology Overview and Case Study.* Available at: https://biochar-us.org/presentation/biochar-stormwater-treatment-technology-overview-and-case-study

Wang, H., Wang, S., and Wang, X. 2020. Effects of biochar as bedding material on ammonia, dust and greenhouse gas emissions in animal houses—A review. *Journal of Environmental Management* 268: 110720. https://doi.org/10.1016/j.jenvman.2020.110720

Wang, W., Huang, J., Wu, T., Ren, X., and Zhao, X. 2023. Research on the preparation of biochar from waste and its application in environmental remediation. *Water* 15(19): 3387. https://doi.org/10.3390/w15193387

Woolf, D., Amonette, J., Street-Perrott, F., Lehmann, J., and Joseph, S. 2010. Sustainable biochar to mitigate global climate change. *Nature Communications* 1, 56. https://doi.org/10.1038/ncomms1053

Ye, L., Camps-Arbestain, M., Shen, Q., Lehmann, J., Singh, B., and Sabir, M. 2020. Biochar effects on crop yields with and without fertilizer: A meta-analysis of field studies using separate controls. *Soil Use and Management* 36(1): 2–18. https://doi.org/10.1111/sum.12546

Zhang, L., Kumar, A., and Singh, P. 2024. Biochar applications in pharmaceutical pollutant removal: A comprehensive review. *Environmental Science and Technology* 58(12): 5234–5248.

Zhang, Y., He, M., Wang, L., Yan, J., Ma, B., Zhu, X. et al. 2022. Biochar as construction materials for achieving carbon neutrality. *Biochar* 4(1). https://doi.org/10.1007/s42773-022-00182-x

Zhang, Z., Chen, X., Lee, D.J., and Ni, J. 2021. Biochar as an additive for animal production: A review. *Environmental Science and Pollution Research* 28(4): 4459–4475. https://doi.org/10.1007/s11356-020-10980-y

Żukiewicz-Sobczak, W., Latawiec, A., Sobczak, P., Strassburg, B., Plewik, D., and Tokarska-Rodak, M. 2020. Biochars originating from different biomass and pyrolysis process reveal to have different microbial characterization: Implications for practice. *Sustainability* 12(4): 1526. https://doi.org/10.3390/su12041526

Conclusion

From the fertile soils of Terra Preta to today's cutting-edge carbon credit frameworks, biochar has traveled a remarkable journey. What began as an ancient practice of enriching soils through the careful use of organic residues and charred materials has now emerged as a scalable, science-backed solution for some of the greatest challenges of our time—climate change, soil degradation, and food insecurity.

The chapters of this book have shown that biochar is far more than a soil amendment. It is a bridge between ancient wisdom and modern innovation, capable of improving agricultural productivity, supporting sustainable industries, and creating long-term carbon sinks that benefit the entire planet. Yet the true power of biochar is not only in its physical properties but also in the systems that surround it—education, policy, and markets.

To realize biochar's full potential, several guiding insights emerge:

Education Fuels Transformation: Awareness and knowledge-sharing are indispensable for empowering farmers, entrepreneurs, policymakers, and communities.

Policies Create an Enabling Environment: Supportive regulations, clear standards, and fair carbon credit pricing are critical for unlocking the scale biochar needs.

Innovation Drives Cascading Benefits: Whether in agriculture, construction, remediation, or carbon markets, biochar's versatility offers opportunities for both climate impact and economic growth.

The path ahead will not be without obstacle as production costs, adoption barriers, and policy gaps remain. But these are surmountable challenges when met with collaboration, creativity, and persistence. Around the world, pioneering companies, researchers, and communities are already proving what is possible.

DOI: 10.1201/9781003467946-13

Ultimately, biochar represents a convergence of past and future: an ancient practice reshaped for modern sustainability. If scaled with care and vision, it can nourish soils, sustain livelihoods, stabilize our climate, and help societies thrive in balance with nature. The invitation is clear: to embrace biochar not just as a tool, but as a movement—one that links innovation with responsibility, economy with ecology, and humanity with the earth itself.

Acronyms and Glossary

Acronyms

Acronyms		Description
ABC Plan	–	Brazil's Low Carbon Agriculture Program
ANSI	–	American National Standards Institute
ANZBIG	–	Australian/New Zealand Biochar Industry Group
APBC	–	Asia Pacific Biochar Conference
ASTM	–	American Society for Testing and Materials
AWC	–	Available Water Content
BBFs	–	Biochar-Based Fertilizers
BCR	–	biochar for carbon removal
BECCUS	–	Bioenergy with carbon capture and storage or utilization
CAGR	–	Compound Annual Growth Rate
CAPEX	–	Capital Expenditure
CAR	–	Climate Action Reserve
CBEN	–	Circular Bionutrient Economy Network
CCTS	–	Carbon Credit Trading Scheme
CDR	–	Carbon Dioxide Removal
CEC	–	Cation Exchange Capacity
CEC	–	Contaminants of Emerging Concern
CFRP	–	Carbon Fiber-Reinforced Polymer
CNTs	–	Carbon Nanotubes
CO_2	–	Carbon Dioxide
CO_{2e}	–	Carbon Dioxide Equivalent
CRCF	–	Carbon Removal Certification Framework

CRCF	–	Carbon Removals and Carbon Farming
CSI	–	Carbon Standards International
DCFC	–	Direct Carbon Fuel Cell
DEP	–	Diethyl Phthalate
DIN	–	German Institute for Standardization
dMRV	–	Digital Measurement, Reporting, and Verification
EBC	–	European Biochar Certificate
EBI	–	European Biochar Industry Group
ECI	–	Engineering Conferences International
EJ	–	Environmental Justice
EMF	–	Electromagnetic Frequency
EMI	–	Electromagnetic Interference
ENCHAR	–	European Biochar Research Network
EPA	–	Environmental Protection Agency
EU ETS	–	European Union Emissions Trading Scheme
FAO	–	Food and Agriculture Organization
FC	–	Field Capacity
GHG	–	Greenhouse Gas
GWP	–	Global Warming Potential
HABs	–	Harmful Algal Blooms
HDPE	–	High-Density Polyethylene
HVAC	–	Heating, Ventilation, and Air Conditioning
IBBN	–	India BioChar and BioResources Network
IBI	–	International Biochar Initiative
ICHAR	–	Italian Biochar Association
ILBI	–	Latin American Biochar Institute
IPCC	–	Intergovernmental Panel on Climate Change
ISO	–	International Organization for Standardization
LBC France	–	Label Bas Carbone France
LCA	–	Life Cycle Assessment
LULUCF	–	Land Use Change and Forestry
METs	–	Microbial Electrochemical Technologies
MFCs	–	Microbial Fuel Cells
MNRE	–	Ministry of New and Renewable Energy's
MOA	–	Made of Air GmbH

MRT	–	Mean Residence Time
MRV	–	Measurement, Reporting, and Verification
N_2O	–	Nitrous Oxide
NAE	–	Nitrogen Agronomic Efficiency
NBP	–	National Biofuel Policy
NGOs	–	Non-Governmental Organizations
NOx	–	Nitrogen Oxides
NRRI	–	Natural Resources Research Institute
NUE	–	Nitrogen Use Efficiency
OBS	–	Oregon Biochar Solutions
OH	–	Hydroxyl Radicals
OPEX	–	Operational Expenditure
OSA	–	Organic Soil Amendments
PAH	–	Polycyclic Aromatic Hydrocarbon
PAW	–	Plant-Available Water
PDD	–	Project Development Description
PFAS	–	Per- and Polyfluoroalkyl Substances
PFOS	–	Perfluorooctane Sulfonate
PICD	–	Participatory Informational Community Development
PWP	–	Permanent Wilting Point
PyOM	–	Pyrogenic Organic Matter
REACH	–	Authorisation and Restriction of Chemicals
RMC	–	Ready Mixed Concrete
ROI	–	Return on Investment
SDGs	–	Sustainable Development Goals
SIPs	–	Structural Insulated Panels
SOC	–	Soil Organic Carbon
SSA	–	Specific Surface Area
UNEP/GEF	–	United Nations Environment Programme/Global Environment Facility
USBC	–	US Biochar Coalition
USBI	–	U.S. Biochar Initiative
USFS	–	United States Forest Service
Verra	–	Verified Carbon Standard
WFD	–	Waste Framework Directive
WHC	–	Water-Holding Capacity

Glossary

Additionality: In a Biochar Project refers to the principle that GHG emissions reductions or carbon sequestration achieved by the project would not have occurred in the absence of the project.

Anthrosol formation: Refers to the development of human-modified soils through deliberate and sustained human activities over extended periods. These soils are characterized by deep, dark, fertile horizons created by the accumulation of organic matter, charcoal, and cultural materials.

Audit: Detailed review and evaluation process aimed at verifying the project's compliance with established standards, methodologies, and objectives.

Baseline: Is the initial set of conditions that describe the state of the environment, soil, or agricultural system before biochar is applied.

Biochar carbon markets: Refer to the organized trading systems and platforms where biochar-based carbon removal credits are bought and sold, connecting biochar producers with entities seeking to offset their carbon emissions or meet climate commitments.

Biochar: A carbon-rich material produced through the thermal decomposition of organic matter under limited oxygen conditions.

Biochar-based fertilizer: Biochar-based fertilizer is a hybrid product that combines biochar with conventional nutrients including organic or synthetic fertilizer to create enhanced fertilizer formulations.

Biochar-infused concrete: An emerging composite material that incorporates biochar particles into traditional concrete mixtures to enhance both performance characteristics and environmental benefits.

BioDrying system for biosolids: Refers to a biological treatment process that combines aerobic decomposition with controlled airflow to reduce moisture content in sewage sludge (biosolids) while achieving partial stabilization of organic matter.

Carbon cascades: Refer to the interconnected series of carbon flows, transformations, and benefits that occur when carbon-based materials like biochar are integrated into various systems and applications. This concept emphasizes how a single carbon input can generate multiple, sequential benefits across different sectors.

Carbon credit: Also known as a certificate, a carbon credit is an intangible permit representing one tonne of CO_2e emissions that have been avoided, eliminated or reduced from the atmosphere.

Carbon dioxide equivalent (CO_{2e}): The standard unit that measures the relative climate impact of various greenhouse gases based on the impact of one tonne of carbon dioxide (CO_2).

Carbon sequestration: Process of capturing and storing atmospheric carbon dioxide (CO_2) to mitigate climate change.

Compliance carbon markets: Are government-regulated carbon trading systems where participation is mandatory for covered entities, typically large emitters who must meet legally binding emission reduction targets or purchase carbon credits to offset excess emissions.

Crediting period: Is the defined timeframe during which a carbon project can generate and claim carbon credits for its measurable and verified reductions or removals of greenhouse gas (GHG) emissions.

Feedstock: Material subjected to thermochemical processes used to create various products such as biochar, char, biooil, syngas and more. The feedstock for biochar consists of biomasses, used plastics or old tires, and may contain contaminants.

Field capacity: Represents the soil moisture content after gravitational water has drained.

Gasification: A thermochemical transformation consisting of decomposition by heat with slightly more oxygen than for pyrolysis occurring at temperatures above 1000°C. Gasification generally occurs in four stages including drying, pyrolysis, oxidation of pyrolysis gases in the presence of oxygen, followed by reduction in the absence of oxygen.

LCA: Is a systematic methodology for quantifying the environmental impacts of a product, process, or service across its entire life cycle.

Lifecycle analyses: Refers to the comprehensive assessment of the environmental impacts associated with the production, application, and overall use of biochar.

MRV: Is the backbone of any credible carbon project. It ensures that all claimed carbon removals are real, accurately calculated, and not duplicated.

Organic carbon: Biodegradable carbon compounds present in the organic fraction of biochar feedstocks. Biochar feedstocks may contain compounds such as sugars, starches, proteins, fats, cellulose and lignocellulose, which are thermochemically degradable.

Permanent wilting point: Indicates the moisture level at which plants cannot recover from wilting.

Project feasibility: Refers to the comprehensive assessment of technical, economic, environmental, and regulatory viability before implementing a biochar-based carbon removal or offset project.

Pyrogenic organic materials: Carbon-rich substances formed through the thermal decomposition of organic matter under oxygen-limited conditions.

Pyrolysis: Thermochemical process of transforming carbon-containing feedstocks in the absence or near-absence of oxygen between 300 and 1000°C, without flames. During pyrolysis, feedstock compounds are broken down into smaller molecules forming light gases (e.g. CO_2, CH_4), heavy gases and solids.

Soil structure: Refers to the arrangement and organization of soil particles (sand, silt, clay) into aggregates or clumps, along with the pore spaces between them.

Volatile carbon markets: Refer to carbon trading systems characterized by significant price fluctuations, unpredictable demand patterns, and rapid changes in market conditions that create uncertainty for both buyers and sellers.

Voluntary carbon markets: Are private trading platforms where carbon credits are bought and sold on a voluntary basis, driven by corporate sustainability goals, individual climate action, or investment strategies rather than regulatory requirements.

Water holding capacity of soils: Refers to a soil's ability to retain water, essential for plant growth and ecosystem health.

Yield: Ratio of the final mass of a product to initial mass of the feedstock × 100%.

Please note that the definitions vary depending on the person, organization, or country that uses the term. To date, there is no official consensus adopted by the entire industry as far as the nomenclature of the terms described above. These terms are therefore the definitions of this book, and the definitions are simplified for this book.

Index

For Product Safety Concerns and Information please contact our EU representative GPSR@taylorandfrancis.com Taylor & Francis Verlag GmbH, Kaufingerstraße 24, 80331 München, Germany